DATA IN THE CLOUD

A CONCEPTUAL DATA MODEL

FOR

AMAZON DATABASE SERVICES

By
Henry M. Nirsberger

Edited/Reviewed By: Bernard E. Durfee, William Rielly, Jennifer K. Senich, and Adelaide M. Nirsberger.

Cover Design By: Robert Sacca

Version 1.0 ... January, 2020 ... based on Amazon Database and EC2 documentation:

- Amazon Relational Database Service, API Reference (API Version 2014-10-31)
- Amazon Aurora, User Guide for Aurora (API Version 2014-10-31)
- Amazon Neptune, User Guide (API Version 2017-11-29)
- Amazon DocumentDB, Developer Guide (API version: 2014-10-31)
- Amazon Redshift, API Reference (API Version 2012-12-01)
- Amazon DynamoDB, Developer Guide (API Version 2012-08-10)
- Amazon ElastiCache, API Reference (API Version 2015-02-02)
- Amazon ElastiCache for Redis, ElastiCache for Redis User Guide (API Version 2015-02-02)
- Amazon ElastiCache, ElastiCache for Memcached User Guide (API Version 2015-02-02)
- Amazon Elastic Compute Cloud User Guide for Linux Instances (2016)
- Amazon Elastic Compute Cloud, API Reference (API Version 2016-11-15)
- Amazon Quantum Ledger Database (Amazon QLDB): Developer Guide (API version: 2019-01-02, Latest documentation update: September 10, 2019)

Preface

Why Is This Book Needed?

Once upon a time, during those thrilling yesteryears of information technology history (circa the late 1960s and early 70s, when "Information Technology" was called "Electronic Data Processing" abbreviated "EDP" -- not "IT"), an IBM computer scientist, Edgar F. "Ted" Codd, conceived and published his theories on the *relational model* for database management.[1] Upon his pioneering shoulders have stood many IT professionals, authors and entrepreneurs, who consequently evolved civilization's contemporary relational database management systems (RDBMS) along with a thriving, competitive, multi-billion dollar worldwide marketplace. Among them were Raymond F. Boyce and Donald D. Chamberlin (also IBM computer scientists) who co-developed SQL (the Structured Query Language) which itself has evolved to become the most familiar and frequently used language for accessing and manipulating RDBMS data.[2,3]

Many years of research, product innovations, versions and releases have blessed humanity with a variety of robust, modern-day, commercially available, SQL-based relational database products (e.g., Oracle DB, Microsoft SQL Server, IBM Db2, etc.). Relational databases are now the foundation of many mission critical, enterprise scale applications such as ERP (Enterprise Resource Planning), CRM (Customer Relationship Management), and SCM (Supply Chain Management) systems -- driving online transaction processing (OLTP), analytical processing (OLAP) and other operational databases for countless customers world-wide.

And you -- the curious but skeptical reader -- are likely a veteran IT *data professional* possibly with substantial relational database background and experience. Your successful career, having yielded uninterrupted paychecks, has focused on the *mainstream* of IT with its apparent fixation on relational databases -- and likely you've productively contributed to, and lucratively benefitted from, a variety of IT projects, such as:

- Implementation of commercial applications using an RDBMS
- Development of custom OLTP applications using SQL for accessing an RDBMS
- Implementation of OLAP, Business Intelligence (BI), or Data Warehouse (DW) applications using an RDBMS
- Administering an RDBMS -- or managing or collaborating with RDBMS database administrators (DBAs)
- Development of data models or database designs for RDBMS applications

You, an unrelenting student of IT, owe much of your professional success to your unquenchable thirst for knowledge. You've acquired and maintained your knowledge of many of the above mainstream subject areas. Powered by the confidence of that knowledge, you have become the "go to" gal or guy for pontificating opinions on many data management challenges facing your enterprise. As a result, perhaps you are (or will someday ascend as) the CDO -- the Chief Data Officer -- for your enterprise, flaunting your RDBMS subject matter expertise. The data that defines the business world, from your perspective (IT life as you know it), rests by and large on *relational database technology,* your professional and technical comfort zone.

But your conscience, that inner silent voice, cautions … *"You can never rest on your knowledge laurels and stop studying IT"* … because …

"Once you stop learning, you start dying" (Albert Einstein)

and …

"If you rest, you rust" (Helen Hayes).

Inspired by a growing number of interactions with other IT professionals …

- Colleagues returning from cloud conferences with unorthodox ideas about storing data as documents without table structures,
- App developers (young, recent-hire upstarts … whippersnappers) hyping MongoDB and other (gasp!) NoSQL solutions,
- IT friends employed at other enterprises … spouting about their forced migrations to the cloud and their acquaintance with new cloud database services,

… you begin to suspect that times are changing for the relational status quo … that there's now much beyond IT *data life* than relational stuff. They toss around numerous, non-relational concepts, buzzwords, and acronyms:

Graph databases, document databases, NoSQL query languages, GQLs, in-memory databases, cache server clusters, cloud database platform services, MPP, immutable ledger databases, blockchain technology, etc.

You are also overhearing scuttlebutt about new use cases and applications that appeal to non-relational solutions, and that new non-relational cloud data services are now ready for prime time … ready for serious consideration:

- Graph databases are now commonplace for building *Recommender Engines* for social media and on-line retail suggestions. The gossip? Such apps with "highly connected" data sets and complex bill-of-material relationships have much better response times and are easier to implement with graph databases and their allegedly more intuitive NoSQL graph query languages.
- New cloud database services deliver astonishingly fast sub-second latency for highly available, global apps (e.g., online retail, gaming, etc.) and readily support tens of thousands of concurrent users -- using NoSQL in-memory databases, data partitioning, cache clusters, and columnar storage.
- "Agile" web app developers are lobbying for a shift to document databases to facilitate app development … "more intuitive and more scalable" they say, than complex relational multi-table joins.
- C-level executives (CxOs) with cyber-security nightmares about threats to data confidentiality, integrity and availability -- espionage of IP, PII, PCI, and HIPAA data … ransomware … denial of service attacks … fraud … sabotage. Can the

implied immutability of new ledger databases and blockchain technology mitigate such threats?

No worries … change is a good thing … you're optimistic! You've grown (and not groaned) through many IT paradigm shifts throughout your career (i.e., many OS, ERP, and RDBMS mergers, acquisitions, and migrations). Now a new paradigm … a new challenge …

from Relational to Non-Relational!

You envision that these new concepts and use cases can be exploited as opportunities to amass additional IT knowledge, wield intellectual power, speak with professional authority beyond your RDBMS comfort zone … and you can emerge as a renowned non-relational hero! So as always, once again, you will be open-minded to these non-relational approaches now available in the cloud.

Fortunately, your enterprise is likely already familiar with moving apps to the Amazon cloud (AWS), having either begun the planning process or actually transitioned pieces of your apps to the cloud (i.e., the low hanging fruit -- the web servers and app servers that "reach back" to databases in your private data centers). But now your insatiable CxOs want more cost savings and additional data center *footprint* reduction. They now want your database servers moved to the cloud! They also want to understand these new non-relational database services as well! Why? Because of new "business imperatives" for competitive advantage: Faster response times for a new global shopping cart app, developing prototype recommender systems, faster time-to-market for apps development, cyber-security and data immutability … blah, blah, blah. And they want you to deliver presentations, plans and proposals for all of the above … pronto!

So you, like many IT professionals world-wide, are currently confronted with the challenge of quickly absorbing basic concepts surrounding the ever growing variety of AWS database services. You will need to do some cramming! As before, because of traditionally tight training budgets, you once again proactively google "AWS documentation" this time for unravelling AWS database services. You are shocked by many user guides, developer guides, and programming reference documents -- over 20 documents -- covering an impressive array of services. You hunger for a high-level summary -- with diagrams and pictures worth thousands of words -- that you can rapidly reuse for promptly producing your presentation. Alas, you quickly conclude that none such synopsis exists.

Introductory videos stun you with the robust collection of AWS database services offering many options for migrating existing relational databases, creating non-relational databases, and many new data concepts beyond your relational security blanket. You also astutely absorb from the user guides that AWS database services are described by a plush "Tower-of-Babel" vocabulary of many new fundamental concepts -- far deeper than water cooler and happy-hour discussions with IT colleagues:

> *Cluster Nodes, DB Instances, Event Subscriptions, Option Groups, Parameter Groups, Replication Groups, Node Groups, Ledger Databases, Merkle Trees, etc.*

Along with basic cloud infrastructure concepts:

> *Regions, Availability Zones, Accounts, Virtual Private Clouds, Subnets, Instances, Security Groups, Subnet Groups, Resources, etc.*

Their esoteric terminology and concepts are inextricably intertwined by many relationships and described by thousands of pages of detailed, technical prose, supported by many intuitive diagrams. Yet, nowhere, within this vast compendium of documentation can you, a veteran shepherd of data, find the one thing that would allow you to develop an instinct for how all of these pieces fit together. You suspiciously wonder …

Is there no *data model*?

What's a data model? As a data professional, you understand that a data model is primarily used for designing databases. But a *conceptual data model* can serve as a mind map for visually understanding basic concepts (entities) and relationships between entities for any complex subject area -- not only a database. Your data modeling intuition kicks in -- you pull out your trusty yellow lined legal pad and you begin to doodle these newly discovered AWS database concepts as entities. Separate boxes indicate Clusters, DB Instances, Parameter Groups, Parameters, Option Groups, Replication Groups, Node Groups, and many more concepts -- drawing lines in between to represent the complex relationships connecting these entities. But then, you ponder to yourself …

- "Where's the data model for each service?"
- "Is there a summary data model comparing and contrasting all of the various AWS database services?"

You quickly skim through all of the user guides. No data model found! You peruse other related docs … again none found. How about general web searches … image searches … again none found! Sadly, it appears that no persuasive, compelling data model has yet been proposed or published for AWS database services. You will need to slog through thousands of pages. You'd rather tolerate a root canal … a colonoscopy … or a frontal lobotomy!

Unmistakably, your IT instincts suggest to you that the cycle time for achieving a confident overall understanding of AWS database services can be substantially squeezed with the help of a conceptual data model summarizing their many concepts and interrelationships. A concise data model can easily clarify your confusion, dissipate the daze, and diminish the duration for coming up to speed with AWS database services.

So here, to the best of this author's knowledge, is likely the first attempt at publishing a conceptual data model for AWS database services. Let this be a testimony to the hope and trust that data modeling can possibly serve as a fresh approach for unravelling the knots of any obscure, nebulous, and highly technical subject area. The venture of attempting to quickly comprehend and contrast AWS database services is indeed one such thorny challenge. Herein you will enjoy numerous, and hopefully intuitive, entity-relationship diagrams; the result of an earnest effort to demystify the complexities of AWS database services.

This endeavor is a consequence of reverse engineering and visually interpreting the narrative found within the library of AWS Database and EC2 documentation,[4] at the time of this publication, specifically:

- *Amazon Relational Database Service, API Reference* (API Version 2014-10-31)
- *Amazon Aurora, User Guide for Aurora* (API Version 2014-10-31)
- *Amazon Neptune, User Guide* (API Version 2017-11-29)
- *Amazon DocumentDB, Developer Guide* (API version: 2014-10-31)
- *Amazon Redshift, API Reference* (API Version 2012-12-01)
- *Amazon DynamoDB, Developer Guide* (API Version 2012-08-10)
- *Amazon ElastiCache, API Reference* (API Version 2015-02-02)
- *Amazon ElastiCache for Redis, ElastiCache for Redis User Guide* (API Version 2015-02-02)
- *Amazon ElastiCache, ElastiCache for Memcached User Guide* (API Version 2015-02-02)
- *Amazon Elastic Compute Cloud User Guide for Linux Instances* (2016)
- *Amazon Elastic Compute Cloud, API Reference* (API Version 2016-11-15)
- *Amazon Quantum Ledger Database (Amazon QLDB): Developer Guide* (API version: 2019-01-02, Latest documentation update: September 10, 2019)

Your feedback, suggestions, and improvement concepts, via social media or otherwise, are eagerly welcomed; and in conjunction with any major updates to the aforementioned references combined with God's willingness, will serve as the basis for future editions.

Who Should Read This Book?

Many IT professionals from various walks of IT life will benefit from the content herein:

- **Developers of Relational Applications and Services --** whose comfort zone is design, implementation, deployment and maintenance of applications implemented using SQL-based, relational database management systems -- and who are now faced with the challenge of migrating existing relational databases to the cloud or creating new relational databases in the cloud -- and need to quickly grasp the basics of AWS database services providing relational database options.

- **Developers of Non-Relational Applications and Services --** whose comfort zone is design, implementation, deployment and maintenance of applications implemented using non-relational approaches (NoSQL, graph databases, document databases, key-value stores, in-memory databases, in-memory caching) -- and who are now faced with the challenge of migrating their applications to the cloud or creating new non-relational databases in the cloud -- and need to quickly comprehend the basics of AWS database services providing non-relational database options.

- **Enterprise Architects** -- especially Data Architects but also Application Architects, Technology Architects, and Business Architects who have been challenged by the prospect of database cloud migration, design of new relational or non-relational applications (using NoSQL, graph databases, document databases, in-memory databases, in-memory caching) and who will be called upon for technical presentations, opinions, demos, or participation in a proof-of-concept.

- **Database Educators/Instructors/Teachers** -- especially those anticipating a need for developing and delivering courseware focused on cloud-based database services. The content herein provides a unique and visual approach as a jump-start launching point for inspiring and quickly enabling presentation material for introducing, summarizing, comparing, and contrasting contemporary cloud-based data services.

- **Data Management Professionals** interested in learning how to apply conceptual data modeling techniques as an approach to describing any highly complex, technical subject matter, not just for describing or designing familiar, commonplace information systems.

- **IT Professionals Acting in Support Roles** -- other than IT applications and services engineering (e.g., operations, application orchestration, continuous integration and delivery, etc.), who will need to support migrated cloud apps and databases and, therefore, also need to quickly absorb AWS database services concepts.

- **Any IT professional Seeking an Introductory Data Modeling Tutorial** explaining diagram syntax and semantics, along with many robust examples of data modeling diagrams.

- **Database Administrators (DBAs)** with limited conceptual data modeling experience, appreciation or confidence, who wish to expand their skills and responsibilities to exploit conceptual data modeling as a method for formalizing requirements, project scoping, and effective communication among all project stakeholders.

- **Any IT professional** who is currently or formerly conversant with one or more AWS database services but would like to review these conceptual models for rapidly refreshing, comparing and contrasting their comprehension of AWS database services.

- **Anyone** with curiosity for considering an example of how a conceptual data model can be used to graphically portray information requirements and concept structure of virtually any perplexing discipline or area of interest, e.g., Cyber-security, Machine Learning, Internet of Things, Blockchain technology, Linear Algebra, Particle Physics, Organic Chemistry, ad nauseam.

What You Will Find

This document will take you deep down the "rabbit hole" into the wonderland of Amazon Database Services -- for a tour of their fundamental concepts and related details, visually supplemented by many intuitive data modeling diagrams. Here's a brief summary of what you will find:

- **Introduction:** The tour begins by providing answers to elementary but crucial questions: What are the Amazon Database Services and Amazon EC2? What is a Data Model? What is a Conceptual Data Model? You will appreciate that AWS database Services are cloud platform services (Platform as a Service -- PaaS). Amazon EC2 provides *Infrastructure as a Service* (IaaS, the foundation upon which all of the Amazon Database Services are built) whereby a customer account dynamically provisions infrastructure components (e.g., virtual servers, firewalls, etc.). A *data model* can be used for visually describing any subject area's essential concepts and relationships. The state-of-the-practice for accomplishing such is to create a specific type of data model, commonly called a *conceptual data model*, which provides a high-level understanding of essential concepts and relationships. Chapters 2 through 9 will provide a conceptual data model for each of the Amazon Database Services -- illustrating basic and critical *entities* (i.e., essential concepts) included on many *entity-relationship diagrams* (ERDs). Created using the ER/Studio Data Architect tool (a product of IDERA, Inc.), ERDs depict a common data modeling diagramming style known as *Information Engineering* (IE) notation.

- **Chapter 1 – Amazon EC2 Basics:** Primitive EC2 concepts (Regions, Availability Zones, Images, Accounts, Instances, Instance Types, VPCs, Subnets, Security Groups, VPC Instances, Classic Instances) are defined and illustrated via simple ERDs. These initial ERDs serve as the basis for providing a tutorial on data modeling concepts and Information Engineering notation: entities, relationships, attributes, Crow's Feet Notation, one-to-many relationships, many-to-many relationships, dependent entities, optional vs. mandatory relationships, identifying vs. non-identifying relationships, foreign keys, primary keys, super-types and sub-types.[5]

- **Chapter 2 – Amazon Relational Database Service (RDS):** The deep dive into AWS database services concepts begins with RDS, a managed web service for implementing familiar relational database management systems (i.e., Oracle DB, Microsoft SQL Server, MySQL, MariaDB, and PostgreSQL) … including definitions and supporting ERDs illustrating concepts such as Database Instances, Option Groups, Options, Event Notification, Reserved Database Instances, Database Backups, Database Logs and Log Types.

- **Chapter 3 – Amazon Aurora:** Aurora is a managed service supporting clusters of database servers for familiar open source RDBMS engines (MySQL and PostgreSQL) … including definitions and supporting ERDs illustrating concepts such as Aurora clusters, a primary DB instance, Read Replica database instances, virtual cluster volumes, DB cluster snapshots, cross-region DB clusters, backtracking, and Serverless clusters.

- **Chapter 4 – Amazon Neptune:** Amazon Neptune is a managed AWS database service supporting graph database engines accessed by NoSQL graph query languages (GQLs). Other topics include graph data structures, highly connected datasets, Gremlin and SPARQL GQLs, recommender engines, labelled property graphs, vertices, edges, and properties. Supporting ERDs illustrate concepts such as Neptune clusters compared and contrasted with Aurora clusters.

- **Chapter 5 – Amazon DocumentDB:** DocumentDB is another managed database service offered by AWS, providing clusters of document database servers and compatibility with MongoDB. DocumentDB is also categorized as a NoSQL database service. Basic document concepts (e.g., collections, documents, fields, embedded documents, etc.) are introduced along with MongoDB operations for creating, reading, updating, and deleting documents. Supporting ERDs illustrate DocumentDB clusters in contrast with Neptune and Aurora clusters.

- **Chapter 6 – Amazon Redshift:** Redshift, based on PostgreSQL, is another managed data service offered by AWS, and targets applications requiring efficient access to very large data sets (e.g., BIDW, OLAP, ETL, etc.). Redshift also supports clusters of database nodes (rather than DB instances). Redshift basic concepts -- leader nodes, compute nodes, massively parallel processing (MPP), partitioned data sets, slices, columnar storage, table restore requests, etc. -- are described. ERDs illustrate Redshift clusters contrasted with DocumentDB, Neptune and Aurora clusters.

- **Chapter 7 – Amazon DynamoDB:** DynamoDB is another NoSQL, managed AWS database service supporting structured and semi-structured data stored as key-value pairs and documents (JSON files). DynamoDB is *serverless* and ideally for implementing global, internet scale apps (e.g., shopping carts, gaming, bidding, etc.). DynamoDB basic concepts are described: tables, global tables, global/local indexes, auto scaling, streams, DynamoDB Accelerator (DAX), cache clusters, caching strategies, DAX clusters, DAX nodes, etc., along with supporting ERDs.

- **Chapter 8 – Amazon ElastiCache:** ElastiCache is another AWS managed caching service for implementing scalable clusters of cache nodes supporting both Memcached and Redis cache engines. ElastiCache basic concepts are described (e.g., caching strategies, lazy loading, data structure servers, Redis commands, Memcached commands, ElastiCache clusters, Redis standalone clusters, Redis replication groups, snapshots, reserved instances, node type parameters, etc.) along with supporting ERDs.

- **Chapter 9 – Amazon Quantum Ledger Database (QLDB):** QLDB, a managed AWS database service, is intriguingly different from previously described AWS database services. QLDB focuses on apps requiring a *ledger* database, a repository for System of Record (SOR) apps requiring complete transaction history. A QLDB ledger blends a smorgasbord of concepts from relational, document, and blockchain paradigms. The following QLDB basic concepts, summarized with supporting ERDs, are described: SQL access to tables of documents (via PartiQL, a SQL-like language); Amazon Ion format (a superset of JSON); QLDB's built-in *journal* of change history; Blockchain concepts (i.e., the journal is an append only, immutable chain of blocks); Verifiability of data integrity (i.e., via SHA-256 hash codes, Merkle Trees, and Merkle Audit Proofs).

- **Appendix**: Summary ERDs for each of Chapters 1 through 9 are replicated, as a matter of convenience, enabling periodic and relatively rapid review of AWS database services and EC2 essential concepts and relationships.

About The Author

Henry M. Nirsberger, also the author of *A Conceptual Data Model for Amazon EC2,* earned an MS in Systems & Information Science at Syracuse University and BA in Mathematics, Le Moyne College. His professional IT career spans five decades starting as a software engineer with Honeywell Information System in 1974, continuing with GE in 1979. While at GE, he assumed a variety of technical, project management, and management positions, culminating as a Senior Architect for GE's Corporate Enterprise Architecture Team, having gratefully contributed to hundreds of IT systems and applications development projects. Currently the CEO of HMN Consulting LLC, specializing in enterprise architecture and data management consulting services, his past professional certifications include CDMP (DAMA), CBIP (TDWI), CFPIM (APICS 1984–2003) and TOGAF 9. As a trained facilitator (MGR Consulting), he facilitated over 600 IT design and planning sessions for data modeling, process modeling, database design, project planning, process improvement, requirements consensus and team building. Presently, he remains a student of data modeling, cloud computing, enterprise architecture, cyber-security, and all aspects of data management.

Acknowledgments

The author gratefully appreciates and acknowledges the efforts of the editors and reviewers of this endeavor. Editing services were adeptly delivered by Bernard E. Durfee (Lambda School), William Rielly (Principal Analytics Engineer, CDPHP), Jennifer K. Senich (VP Operations, Ithos), and Addie Nirsberger (the author's perfect 10 wife, best friend, and soul mate). The author also salutes Bernard E. Durfee for inspiration and ideas used for portions of narrative in the introduction chapter. The book cover was designed by Bob Sacca, one of the author's all-time favorite GE managers. The author is also grateful to Joy Ruff (Product Marketing Manager, IDERA) for exemplary customer support and providing an evaluation license of IDERA ER/Studio Data Architect from which all entity-relationship diagrams herein were created. The author also recognizes the outstanding support and infinite patience demonstrated by the Amazon AWS Support Center, whose technical support staff rapidly responded and proficiently resolved 79 cases for questions on all Amazon database services raised by the author over a period of several months.

Dedication

This effort is dedicated to Mom and Dad … Anna and Michael … intrepid immigrants to America … in search of freedom and opportunity. The memory of your love, laughter, values, determination, and encouragement for continuous learning inspire me each day.

Henry M. Nirsberger
Clifton Park, New York, USA
Henry.Nirsberger@gmail.com

Table of Contents

Preface

Why Is This Book Needed? .... *iii*

Who Should Read This Book? .... *vii*

What You Will Find .... *viii*

About The Author .... *xi*

Acknowledgments .... *xi*

Dedication .... *xi*

Introduction **1**

What are the Amazon Data Services? *7*

What is a Data Model? *12*

What is a Conceptual Data Model? *13*

Chapter 1 – Amazon EC2 Basics **17**

Regions, Availability Zones, Images, Accounts *17*

Data Modeling Tutorial ... Basics *18*

Instances and Instance Types *23*

VPCs, Subnets, IP Address Ranges, Security Groups *24*

Instances ... VPC vs. Classic *27*

Resources, Tags *30*

Chapter Summary *32*

Chapter 2 – Amazon Relational Database Service (RDS) **35**

Amazon RDS Introduction *35*

Option Groups and Options *39*

Event Notification *41*

Reserved Database Instances *43*

Database Backups *46*

Amazon RDS Database Logs and Log Types *48*

Chapter Summary *50*

Chapter 3 – Amazon Aurora **55**

Amazon Aurora Introduction *55*

DB Clusters *57*

Aurora Clusters *59*

Chapter Summary *63*

Chapter 4 – Amazon Neptune **65**

Amazon Neptune Introduction ... 65

Gremlin Introduction ... 67

SPARQL Introduction ... 69

Neptune Clusters ... 73

Neptune vs. Aurora Cluster Differences ... 76

Chapter Summary ... 80

Chapter 5 – Amazon DocumentDB **83**

Amazon DocumentDB Introduction ... 83

DocumentDB Clusters ... 85

DocumentDB, Neptune and Aurora Cluster Comparison ... 88

Chapter Summary ... 93

Chapter 6 – Amazon Redshift **97**

Amazon Redshift Introduction ... 97

Redshift Clusters ... 99

Redshift Table Restore Requests ... 103

Redshift, DocumentDB, Neptune and Aurora Cluster Comparison ... 104

Chapter Summary ... 106

Chapter 7 – Amazon DynamoDB **109**

Amazon DynamoDB Introduction ... 109

DynamoDB Basic Concepts ... 111

Global Tables ... 113

Global and Local Indexes ... 114

Auto Scaling ... 116

DynamoDB Streams ... 118

DynamoDB Accelerator (DAX) ... 120

DAX Clusters ... 123

DAX Nodes ... 124

Chapter Summary ... 126

Chapter 8 – Amazon ElastiCache **129**

Amazon ElastiCache Introduction ... 129

Redis Introduction ... 131

Memcached Introduction ... 133

ElastiCache Clusters ... 134

ElastiCache Replication Groups ... 138

ElastiCache Backups ... 140

ElastiCache Reserved Instances, Node Type Parameters ... 142

Chapter Summary *144*
Chapter 9 – Amazon Quantum Ledger Database **147**
Amazon QLDB Introduction *147*
SQL-based Access to Document Data *148*
Built-in History and Change Data Capture *149*
Immutability *150*
Verifiability *151*
QLDB Basic Entities and Relationships *154*
QLDB Journal Entities and Relationships *155*
QLDB Digest Entities and Relationships *158*
Chapter Summary *159*
Appendix – Summary of Summary Diagrams **163**
Chapter References, Notes, Suggested Reading **175**

Introduction

You, the inquisitive and intrepid reader with a voracious appetite for knowledge, are most likely an IT professional -- perhaps an accomplished, veteran IT professional. Perhaps ... better yet ... you are an IT *data* professional of some sort. Or at least you purport to be, or are reputed to be, a data professional for whatever enterprise you are currently employed.

For a number of years -- throughout your illustrious career -- you've likely enjoyed uninterrupted paychecks, steady salary increases, possibly a number of increasingly responsible job promotions, and job continuity within and across multiple employers. You appreciate the benefits that your IT career has provided you (and yours), and you are thankful. For the most part, your career has dwelled in the *mainstream* of the IT data world. The majority of your data management experience -- the evolution of your career -- likely included (or perhaps will include if you are an aspiring IT professional) encounters and familiarity with many of the following:

- Relational database management systems (RDBMS) -- products such as Oracle DB, MySQL, Microsoft SQL Server, IBM Db2, PostgreSQL, etc.
- Many application development projects using relational databases and SQL.
- Many OLAP (Online Analytical Processing) and ROLAP (Relational OLAP) Business Intelligence (BI) and Data Warehouse (DW) implementations.
- Implementation of commercial applications (e.g., ERP, CRM, and SCM products) using an RDBMS.
- You've developed, reviewed and/or approved many data models and database designs.
- You've rescued engineering teams by troubleshooting RDBMS performance issues related to poor response times or throughput.
- You've collaborated closely with RDBMS database administrators (DBAs), or ...
- You yourself are (or have been) an accomplished DBA or DBA leader.

You owe much of your professional success to your constant craving to acquire IT knowledge and to effectively apply that knowledge to many of the above mainstream, data-related scenarios. It's also likely that you are largely self-taught, because training and T&L budgets have typically been tight. They -- your CxOs or engineering leaders -- frequently come to you (hat in hand, often humbled and uncertain about new or evolving IT concepts and related vendor products) for advice, counsel and recommendations about assorted data management issues and strategies: data governance, data architecture, BIDW, master data management, metadata repositories, data modeling tools, database performance, etc. You've reliably served as the "go to" gal or guy for guidance on many data questions and challenges within your enterprise.

But now ... you are currently employed by the Super Big Mega Corporation (SBMC), which is a wholly owned subsidiary of the Ultra Super Big Mega Corporation (USBMC). You have recently been promoted (or perhaps wish someday soon to be promoted) to be the Chief Data Officer (CDO) of SBMC reporting to the CIO, who reports to the CEO. SBMC has grown in recent years through many mergers and acquisitions. As the CDO of SBMC, you are (or hope to be) responsible for all-things-data, including several hundred heterogeneous, relational databases, all hosted within SBMC's private corporate data centers.

But wait a minute! You can never rest on your IT knowledge laurels and stop learning. You continue to appreciate and understand that the quest for IT knowledge is perpetual. To your credit, you've always been an enthusiastic, unrelenting student of IT. Perhaps you can relax a bit on the "T" in "IT" -- the technology -- the technical infrastructure underlying and supporting SBMC's data repositories. The servers, storage and network components are ever evolving and are rapidly becoming an embryonic commodity in the cloud -- perhaps portending fewer career opportunities for technology professionals unless employed by a cloud provider.

Your inner stream of consciousness, as you prepare to undertake this new CDO career challenge, encourages and inspires you …

> *"Yes ... you are a well-reputed, accomplished data professional ... an RDBMS leader and subject matter expert ... a worthy and plausible CDO."*

… but, at the same time, it cautions you …

> *"Yet ... there are many blind spots in your IT knowledge beyond your RDBMS comfort zone. Are you really ready for these new CDO responsibilities?"*

You are aware of these blind spots, and you know you are (as of yet) ill-equipped to speak with the confidence and authority expected of a well-informed CDO. These blind spots are lurking behind many new buzzwords and acronyms:

- NoSQL
- Graph database engines and Graph Query Languages (GQLs)
- Document databases and database engines
- In-memory database engines with cache clusters
- Cloud database platform services … IaaS and PaaS
- Big Data
- Machine Learning
- The Internet of Things
- Ledger databases and Blockchain Technology

When BigData, HADOOP, and MapReduce emerged as breaking news, your CIO (embarrassed because she or he couldn't crisply explain the concepts, benefits and use cases to curious C-level colleagues in a board meeting) begged you to bring them up to speed with a quick presentation. Unfortunately, you were also, at that time, less than informed about these new technologies … and you had to swiftly scramble (self-taught via google) to deliver a plausible presentation.

Indeed, RDBMSs have been your comfort zone. You've vaguely heard about but never seriously had to unravel alternative approaches for replacing or supplementing relational solutions. To survive and flourish professionally, you've never had to seriously consider alternatives. You've been a loyalist to (and a staunch supporter of) the relational database world. You know and appreciate the richness and power of SQL. You've conquered convoluted SQL subqueries, obscure bill-of-material relationships (many-to-many, self-referential), correlated queries, multi-table inner/outer joins, etc. You've discovered and corrected Cartesian products and many other SQL complexities, yielding an abundance of management awards. And -- let's face it -- pretty much all of the mission-critical ERP, CRM, and SCM systems encountered in your career are relational. So, why bother with alternative database approaches?

Why? Because your peers, colleagues, marketing and sales folks (familiar with SBMC competitor systems), recent-hire app developers (*relational heretics*), and CxOs are starting to whisper about these alternative approaches, all of which are available in the cloud! Henry Ford was woefully behind the times in stubbornly adhering to the Model-T (reluctantly and reactively responding to the competitive variety and colors pioneered by Alfred Sloan at General Motors). Could the same be true for relational die-hards? A bold new world of use cases and cloud data products -- beyond the relational realm -- are now ready for serious consideration:

- *Recommender engines* (using graph databases and *graph query languages*) are now commonplace for making social media recommendations (using friendly, familial, or professional connections) and on-line retail recommendations (using connections between related products/services purchased, or destinations and attractions visited at the same time or by customers in the same vicinity).[6] SBMC has yet to implement a proposed recommender engine for its global online retail system. The gossip is that a traditional relational database would not be nearly as effective as a graph database engine.
- A new breed of "agile" Web app developers (SBMC recent hires) -- overheard during water cooler discussions -- have been lobbying for migration to document databases (e.g., MongoDB) to facilitate app development. Why? Because they are comfy dealing with JSON (JavaScript Object Notation) files, and they say "it's more intuitive" for programmers to store and manipulate document data as JSON files rather than being hampered by normalizing data into relational tables and executing complex multi-table joins.
- Marketing folks are groaning over reports of dwindling market share. SBMC's competitors have all implemented online retail systems boasting of astoundingly fast, millisecond response times (and of surprisingly consistent results for their customers globally) in comparison to SBMC's not-so-snappy legacy relational retail system. SBMC's CTO has discovered (while attending a recent AWS re:Invent conference) that the secret sauce being developed by the competition is based on NoSQL *in-memory* databases and partitioning data in a *cache cluster* to achieve order-of-magnitude latency improvements.
- Cyber-security anxieties are also keeping the SBMC CxOs awake at night. SBMC competitors have been rocked by damaging data confidentiality, integrity and availability issues -- espionage of product designs and intellectual property (IP), personal identity (PII), and payment card (PCI) data … ransomware and denial of service (DoS) attacks … insider threats leading to fraud. No doubt, SBMC CxOs worry that criminal conspiracies may already have been hatched both outside and within SBMC! They wonder: Can a ledger database using blockchain technology mitigate the ever growing number of threats to data security?

You have successfully endured and prospered throughout many momentous IT paradigm shifts in your career:

- Hardware and OS vendor mergers and acquisitions (M&A)
- ERP, CRM and SCM vendor M&A
- Database technology evolution from hierarchical and network (CODASYL) database management approaches to relational databases
- RDBMS vendor M&A and migration to different vendor RDBMS products

But now, you are confronted with yet another challenge … perhaps a more monumental paradigm shift. There are now many other database management flavors beyond the relational world -- new and different use cases demanding diverse solutions. Perhaps SBMC needs a more open-minded, daring, Alfred Sloan type visionary -- rather than an RDBMS dogmatist to be its new CDO. SBMC needs to seriously consider all of the new and exciting non-relational approaches now available in the cloud. You need to do some cramming!

Fortunately, SBMC has already had tremendous success in moving web and app servers for its low risk apps to the cloud. SBMC is now tinkering with the possibility of moving its low risk, non-mission-critical database servers -- no confidential data -- to the Cloud! By now, you know the routine:

- Act I … SBMC CxOs will first want to migrate the RDBMS low-hanging fruit … kick the tires … make sure that everything works … that latency is acceptable … verify cost savings … minimize "reach back" to existing database servers … and then decommission existing database servers.
- Then they undoubtedly will ask for an Act II … migrating medium risk databases and … ultimately …
- Act III … now that they are comfortable with cloud security and encryption, it is time to prudently migrate mission-critical databases to the cloud … and finally ERP apps will be decommissioned in favor of SaaS.

So then, what to do about non-relational and the cloud? You can no longer ignore these technologies. You cannot, once again, kick the can down the road. You can't fake it. Retirement for you is not yet on the horizon … too many responsibilities … married with children … current or looming college costs. Self-actualization is still but a distant hope, a fleeting dream.

The day has arrived … the edict (i.e., the non-relational and cloud database migration edict -- the proverbial "two-by-four on the forehead") is finally delivered to you from your boss, the CIO, in a manner similar to many past management directives, communicating that this paradigm shift is not simply a "business need" but is in reality an "existential threat" of truly paramount importance. Perhaps you have had (or anticipate that you imminently will have, or know someone who has had) a dialogue with a CIO as follows:

- Your boss, the CIO, sticks his or her head into your office and blurts out …

> ***"Hey … Good News! I just got out of our quarterly board meeting. ________ was also there … we were both proud as peacocks!"***

Fill in the name of your CTO.

- You respond ….

> ***"Great! How'd it go?"***

- Your CIO continues ….

> ***"I presented … they loved it! We finally decommed over 200 web and app servers from our datacenters up into AWS heaven. They thanked us profusely … gave us both more stock options and, get this -- the go-ahead to throw a victory party and give management awards to the cloud migration team!"***

- You respond ….

> ***"Wow! Nice! Congratulations! … yep … Migrating our front-end servers to the cloud was a no brainer … I'm sure they liked shrinking the footprint of our datacenters."***

- Your boss continues ….

> ***"Yep … I also presented the numbers on the CAPEX we saved with pay-by-the-drink, lower storage costs, autoscaling, not paying for idle QA and dev servers … yadda yadda yadda."***

- You respond ….

> ***"Excellent! I betcha they'll next ask to come up with a plan for consolidating our remaining datacenters!"***

- Your boss continues ….

> ***"Well … that's the not-so-good-news … you know, these guys on the board are insatiable … unappeasable. Now they want an Act II … database servers! Now that we have web and app servers in the cloud, they understand 'reach back' from AWS to our database servers. Now they think there's more gold to be mined if we can migrate database servers to live alongside the web and app servers -- at least for the low risk, public data stuff … our Microsoft and Oracle servers … there's no appetite yet for confidential data or proprietary stuff."***

- You respond ….

"I got it … if someday we all get comfy with cloud security, then they'll want Act III … medium risk and mission critical crap moved as well."

- Your boss continues ….

  "Yep … it gets worse than that … they heard a bunch of presentations on AWS database services … more than relational. Stuff I've just begun to hear about … Neptune, DynamoDB, Redis, DocumentDB, in-memory data, cache clusters. Our competitors are using these for recommender systems, insane response times, disaster recovery, global reach … blah blah blah. They want to know more … like how can we leverage these services? When will we have a recommender engine? Why are our response times crappy compared to theirs? So the perception is our competitors are ahead of us … so, yet again, we're behind the eight-ball!"

- You respond ….

  "So … DAAAH! … Of course we need to jump on this as quickly as possible … Any dollars left in the training budget?"

- Your boss continues ….

  "No bucks left in T&L but … yes … we have some training pesos that we need to use by year-end anyway. So you can start looking into scheduling a few on-site training sessions for your team. Go ahead and google 'AWS documentation' … check out their database services. Knock yourself out … they'll want an update to hear our story ASAP."

- You respond ….

  "Yep … deja vu all over again … reminds me of two years ago, when the app teams had to bone up on EC2 with no training budget and no T&L."

- Your boss continues ….

  "Yep … you get it! By the way, thanks again for recommending that EC2 data modeling book by what's his name -- I forgot his name. I used some of his diagrams to explain cloud basics to the board when we launched the Web and App server migration. I hope there's something similar for AWS database services. It would save a lot of time coming up to speed!"

Your guiding principle, mantra and motto (perhaps the foundation of your successful job continuity) has obsequiously always been to "enthusiastically do whatever your boss tells you to do … unless unethical or immoral." So like before -- plunging into Amazon EC2 (Elastic Compute Cloud) -- you once more google "AWS documentation" and begin to peel back the layers of this new onion. Again you are bombarded with a plethora of AWS database services and associated documents: User guides, developer guides, API (Application Program Interface) and CLI (Command Line Interface) reference documents, SDKs (Software Development Kits) -- dozens of documents with thousands of pages covering all of the AWS database services. You fantasize for a single, high-level, concise view that summarizes, compares and contrasts this assortment of AWS database services -- diagrams and pictures worth thousands of words -- that you can rapidly reuse for producing your presentation. But you quickly conclude that no such Rosetta Stone for the cloud exists.

After perusing a few overview web pages and viewing several introductory videos, you are impressed (No -- astonished and intimidated!) by the robust collection of AWS database services. You conclude that learning about these services will indeed be a daunting endeavor … perhaps even your final learning frontier as an IT professional ... then maybe you can peacefully retire! You quickly conclude that there are many options for migrating your existing relational databases and many options for creating new non-relational databases. Your bleary eyes are further besieged by myriad new concepts launching you far beyond your relational comfort zone: Cache Clusters, Neptune, Graph Data Bases, Graph Query Languages, Gremlin, SPARQL, Document databases, MongoDB, In-memory databases, Redis, Memcached, RDS, Aurora, NoSQL, Ledger databases, Merkle Trees, etc.

Hence, you do what you've done so many times in the past when previously confronted by yet another major IT paradigm shift (e.g., demystifying Amazon EC2). Yes, without the benefit of formal training, you voraciously, religiously, studiously and independently embark on pouring through whatever documents, tutorials and online videos are readily available. Soon, you will undoubtedly be asked to either make a presentation, offer an opinion, provide a demo, migrate an existing relational database, or participate in a more complex proof-of-concept (e.g., a graph database for your new recommender engine), etc.

So you need to dive more deeply into all of these AWS database services and, of course, answer the immediate question …

What are the Amazon Data Services?

You discover that AWS offers the following database services:

- Amazon Relational Database Service (RDS)
- Amazon Aurora
- Amazon Neptune
- Amazon DocumentDB
- Amazon Redshift
- Amazon DynamoDB
- Amazon ElastiCache
- Amazon Quantum Ledger Database (QLDB)

Each of these database services has its own set of introductory documents and tutorial media. Documentation for each of these services focuses on its own features, offerings and characteristics … with little elaboration about nuances and distinctions between it and other sibling services.

You quickly conclude from overview summaries in their respective user guides that each service is actually implemented as a *managed service* built on top of Amazon EC2. Thus, EC2 is the foundation, the cloud infrastructure, upon which all AWS database services are implemented. EC2 provides, what is referred to in cloud computing parlance as, *Infrastructure as a Service (IaaS)*. The National Institute of Standards and Technology (NIST) in their NIST Definition of Cloud Computing defines Infrastructure as a Service (IaaS) as follows: [7]

> ***Infrastructure as a Service (IaaS)***. *The capability provided to the consumer is to provision processing, storage, networks, and other fundamental computing resources where the consumer is able to deploy and run arbitrary software, which can include operating systems and applications. The consumer does not manage or control the underlying cloud infrastructure but has control over operating systems, storage, and deployed applications; and possibly limited control of select networking components (e.g., host firewalls).*

IaaS essentially means that an AWS account can be established for a customer (e.g., an individual or corporate enterprise), and each customer can readily and dynamically provision infrastructure components (e.g., virtual servers, network devices, storage devices, etc.). EC2 delivers IaaS by providing robust self-service interfaces, such as a web-based user interface (i.e., the Amazon EC2 console), a CLI, and an API.

AWS database services, as managed services dependent on EC2, are often referred to as cloud *platform* services. Each AWS database service is an example of what is referred to in cloud computing parlance as a *Platform as a Service (PaaS)*. NIST in the same NIST Definition of Cloud Computing defines Platform as a Service (PaaS) as follows:[7]

> ***Platform as a Service (PaaS)***. *The capability provided to the consumer is to deploy onto the cloud infrastructure consumer-created or acquired applications created using programming languages, libraries, services, and tools supported by the provider. The consumer does not manage or control the underlying cloud infrastructure including network, servers, operating systems, or storage, but has control over the deployed applications and possibly configuration settings for the application-hosting environment.*

Essentially, each AWS database service is a cloud platform service providing a framework for greatly simplifying and facilitating database provisioning and ongoing maintenance. PaaS in this case substantially reduces the need for dealing with the underlying database infrastructure (i.e., servers, storage, network, OS, database software, etc.) and related data management issues (i.e., upgrades, patching, capacity scaling, database backups, high availability, etc.). PaaS enables engineers to more readily focus on creating and maintaining their applications -- less distractions from database infrastructure -- fewer maintenance headaches -- faster provisioning -- and possibly fewer DBAs?

The introductory summaries by themselves reveal many new concepts and a substantial new technical vocabulary. This reaffirms your fear that the challenge of achieving a solid understanding of AWS database services will indeed be formidable -- a heavy lift. You yearn for some sort of "Cliffs Notes" version of AWS database services … but are still skeptical that such a handbook exists. You drill down further by swiftly scanning the remaining sections of each user guide, and you begin to discern and appreciate that …

- **Amazon Relational Database Service (RDS)** enables the creation of DB instances (a type of EC2 instance), which are ostensibly virtual machines running a relational database engine. But these DB instances are intertwined and related to a plethora of new vocabulary which you will be called upon to illuminate: Option groups, parameter groups, encryption keys, subnet groups, read replicas, event notifications, event subscriptions, snapshots, log types, etc.

- **Amazon Aurora** enables the creation of *database clusters* consisting of many DB instances (i.e., compatible with MySQL and PostgreSQL, open source relational engines). Each cluster is also related to security groups, a subnet group, a parameter group, an engine version, and is a source for event notifications. But Amazon Aurora is also entangled with an abundance of new vocabulary: Aurora clusters, primary DB instances, read replica DB instances, read replica DB clusters, virtual cluster volumes, cross-region DB clusters, backtracking, change records, serverless DB clusters, etc.

- **Amazon Neptune**, like Aurora, also enables the creation of *database clusters* consisting of many DB instances, but this time for *graph databases* -- a different non-relational animal. Neptune is also intertwined with a glut of new jargon and has several stark differences with RDS and Aurora -- the boggling of the mind continues: NoSQL, graph data structures, vertices, edges, properties, graph query languages (GQLs), Gremlin, SPARQL, recommender engines, labelled property graphs (LPGs), triplestores, RDF reification, etc.

- **Amazon DocumentDB**, like Aurora and Neptune, also enables the creation of *database clusters,* but this time for *document databases* -- yet another NoSQL non-relational beast. You again are inundated with a flood of new lingo and substantial differences with RDS, Aurora, and Neptune: Document databases, MongoDB, semi-structured data, key-value pairs, JSON, collections, documents, embedded documents, etc.

- **Amazon Redshift** -- like Aurora, Neptune and DocumentDB -- Redshift also enables the creation of *database clusters* but this time (a vocabulary shift) consisting of many *nodes* not DB instances. At first, you are lulled into a false sense of security because, with Redshift, you are seemingly back in your comfort zone -- the relational world (PostgreSQL). But the euphoria is short-lived because once again you are slapped in the face with more mumbo jumbo and dazzling differences with RDS, Aurora, Neptune, and DocumentDB: Columnar data, leader nodes, compute nodes, massively parallel processing (MPP), partitioned data sets, slices, cross-region snapshot copies, table restoration, etc.

- **Amazon DynamoDB**, like Neptune and DocumentDB, offers yet another NoSQL approach, but it's totally serverless -- without explicit support of clusters of DB servers. Instead? Clusters of cache servers! You are again humbled -- more blind spots in your IT knowledge and many more new concepts to absorb: Provisioned throughput settings, read capacity units (RCUs), write capacity units (WCUs), auto scaling policies, schemaless, global tables, multimaster database, streams, shards, cache clusters, item cache, query cache, eventually and strongly consistent reads, etc.

- **Amazon ElastiCache** represents yet another significant paradigm shift. ElastiCache (like Neptune, DocumentDB and DynamoDB) offers another NoSQL approach, but without the support of any back-end database servers. Instead … clusters of cache servers are provisioned for in-memory storage of data maintained as a key-value store. But memory is volatile, right? No worries … it's still possible to persist and recover data via ElastiCache backups and change logs. You thought you've seen it all, but now again … many more buzzwords: Memcached, Redis, *lazy loading* and *write through* caching strategies, abstract data types, data structure servers, Base64 encoding, standalone clusters, replication groups, etc.

- **Amazon Quantum Ledger Database (QLDB) …** Yes, you have a warm feeling in your tummy about the relational world, and you are beginning to warm up to document databases. Hopefully, you are at least somewhat aware of blockchain technology, Bitcoin and the forking cryptocurrency controversies. But now QLDB proposes an amalgamation of familiar concepts from each of these three approaches -- the relational, document DB, and blockchain paradigms. QLDB, eye-poppingly different from previously described AWS database services, is basically a three-amigos, quasi relational-document-blockchain approach.

  QLDB focuses on apps requiring a *ledger* database for System of Record (SOR) apps requiring complete transaction history (e.g., eCommerce online order tracking and fulfilment). You know the effort it takes to log history in conventional databases (e.g., triggers, stored procedures, partitioned tables, audit logs, etc.). QLDB waves bye-bye to such effort by providing a built-in, append only *journal* of change history (based on blockchain concepts). The journal is also purportedly immutable and verifiable. This offers hope to C-level executives fearful of cyber-security threats to data integrity. QLDB is yet another challenge to your relational comfort zone with a bunch of new concepts to fathom: SHA-256 hash codes, Merkle Trees, Merkle Audit Proofs, etc.

Many new data concepts … far beyond the relational world ... lots of one-to-many relationships … lots of many-to-many relationships. While studying the many user guides, your data management intuition kicks in. You curiously begin to doodle (on your trusty notepad of course) boxes as entities (data concepts) and relationship lines between entities. Separate boxes are sketched for Database Clusters with DB Instances, Cache Clusters with Cache Nodes, VPCs, Availability Zones, Subnet Groups, Option Groups, Parameter Groups, Event Subscriptions, etc. -- judiciously and artfully connected by lines with crows-feet.

Hmmm … a Database Cluster has many DB Instances … A Cache Cluster has many Cache Nodes … A Replication Group is a special kind of Cache Cluster … Subnet Groups have many Subnets … but a Subnet can belong to many Subnet Groups -- a many-to-many relationship. In essence, you've begun the process of reverse engineering the prose transcribed in the user guides into a (far more useful) visual data model … and at a high level it's all starting to make sense!

You surmise that the robust collection of Amazon database services are an amazing accomplishment with many complex concepts and relationships beyond these basic concepts and simple relationships. And you then ponder to yourself and begin a subconscious discussion …

- "I'm squiggling … doodling a data model."
- "There must be existing data models somewhere among all of these documents."
- "Where's the official data model for each of these services?"
- "Is there a unified model that compares and contrasts these services?"
- "What are the differences between all of these clusters? How do they compare?"
- "Am I reinventing the wheel by scribbling these sketches?"

Skimming through the remainder of each of the user guides shows no evidence of a data model. No data model found! Nothing that even remotely looks like a plausible entity-relationship diagram (ERD). Surely, this must be an anomaly. You are thinking to yourself that other AWS database services docs must have at least a conceptual data model, if not an actual physical database design model. You peruse other data services docs … you search the web … even an image search … again, none found! Sadly, it appears that no persuasive data models have yet been conceived, proposed or published for any of the Amazon database services.

Clearly, your IT instincts suggest to you that the cycle time for achieving an assertive overall understanding of Amazon data services can be substantially compressed with the help of conceptual data models summarizing their many concepts and interrelationships. Supporting ERDs focusing on each database service would undoubtedly be incredibly helpful.

But this author's experience is that many contemporary IT professionals have not had substantial exposure to (and, therefore, have not matured an appreciation for) data modeling concepts, tools and methods historically employed in information systems development when, perhaps, "build" was the predominant IT choice versus "buy." The evolution to present-day commercially available enterprise scale ERP, CRM and SCM systems has boosted "buy" to the forefront as a more palatable business decision over "build." Arguably, it appears that there is currently much less need to exploit data modeling for requirements analysis supporting new applications development. But, it remains a compelling tool for describing the information needs of any complex system or process.

So, just in case you are a less seasoned IT professional … or more a prodigy of the "buy" era … or perhaps somewhat doubtful and not persuaded by data modeling benefits, the immediate question that needs to be answered next is …

What is a Data Model?

Circa the early 1960s and prior to Ted Codd theorizing the relational model, a General Electric software engineer, Charlie Bachman,[8] gave rise to the dawn of data modeling by illustrating boxes with connecting lines to describe data structures for databases implemented via one of the earliest commercially available database management systems. His diagrams were known as Bachman Diagrams, and he also coined use of the term, "database entity." Upon his pioneering shoulders have stood many data management professionals, authors and practitioners, who consequently evolved civilization's contemporary data modeling concepts, graphical languages, methods and tools.[9]

Since then, data modeling diagrams (aka entity-relationship diagrams, ERDs) have effectively been used for designing databases and scoping project boundaries for countless complex information systems. Beyond their initial primary usage for database design, data models have also been effectively applied for formalizing and communicating information requirements for business processes, and for illuminating many convoluted, nebulous subject areas. Data modeling, a veritable mind mapping technique, has amply assisted in making many multifaceted disciplines of study more lucid and logical because pictures are indeed worth thousands of words.

Abundant data modeling literature (text books, eBooks, technical papers, journal articles, conference proceedings, human factors studies, blogs, etc.), from both an academic and practitioner perspective, has been published by many scholars and data management professionals. Collectively, these authors have offered many varying definitions for the concept of a *data model* and associated data modeling activities and tools for creating data models. Graeme Simsion, in his extraordinary compendium entitled *Data Modeling Theory and Practice*, amassed over a thousand references to such publications.[10]

Prior to illustrating sample definitions, we begin with the simple assertion that an information system fundamentally consists of software (i.e., application programs) and a database -- a repository for persisting data that is collected, stored, retrieved, transformed, or otherwise manipulated by application software. Simsion provides copious evidence that many authors propose the viewpoint that an information system's database is the primary focus of -- and is specified by -- a data model. In this case, the perception is that data modeling is essentially a *design activity* in that a data model specifies the various classes of data that will be maintained within the database, and describes how such data will be logically organized and internally structured within the database.

Alternatively, many other practitioners propose definitions which suggest that a data model is primarily a *descriptive activity* rather than a design activity. In this case, the focus is to document requirements in terms of essential concepts (i.e., persons, places, products, services, suppliers, transactions, events, assets, etc.), and relationships between such concepts, for a chosen subject area or subject matter significant to an organization or business.

The *DAMA Guide to The Data Management Body of Knowledge* (henceforth herein referred to as the DAMA-DMBOK), a collaborative effort influenced by approximately one hundred data management professionals, provides practical, down-to-earth definitions for both a data model and data modeling -- definitions blending both perspectives. Simply summarized, a data model is used for defining and describing a subject area's essential concepts and relationships; or a data model is used to analyse and specify a subject area's database design.[11]

Indeed, our primary objective is to facilitate an understanding of Amazon database services and their essential concepts. The state-of-the-practice for accomplishing such is to craft a specific type of data model, commonly called a *conceptual data model*. So ....

What is a Conceptual Data Model?

As a result of these two viewpoints (i.e., data modeling is a design activity vs. a descriptive activity), many data management practitioners acknowledge the existence of three different categories of data models:[12]

- Conceptual Data Models (CDM)
- Logical Data Models (LDM)
- Physical Data Models (PDM)

Briefly, a conceptual data model (CDM) embraces the descriptive point of view by providing a high-level understanding of essential concepts and their requirements. A logical data model (LDM) fills in many of the details intentionally omitted by the CDM. The LDM results in a neutral database design, which is impartial and not aligned with any vendor-specific database management system (DBMS) implementation. A physical data model (PDM), in effect, translates the vendor-agnostic LDM into the actual database design tailored to a specific DBMS such as an Oracle DB, Microsoft SQL Server or IBM Db2 database. The PDM adapts the LDM by taking into consideration constraints imposed by the DBMS. A PDM can also adjust the LDM design for satisfying performance requirements.

Again the DAMA-DMBOK provides a simple, straightforward definition for a conceptual data model and its purpose.[11] To a large extent, a CDM's intent is to visually convey, via ERDs, a summary-level understanding of the semantics of a specific subject area or subset of a subject area (i.e., a lower-level function or process supporting a subject area). Subject areas common to most enterprises include Research and Development, Engineering, Production, Marketing, Sales, Logistics, Human Resources, Finance, Information Technology, etc., for which generic data models exist. A CDM illustrates a subject area's essential concepts -- commonly referred to in data modeling parlance as *entities* -- fundamental to understanding the subject area at a rudimentary level. A CDM's entities reflect essential concepts arising from business vocabulary, terms, and jargon pertinent to the CDM's assigned subject area. A CDM does not apologize for omitting low-level detail. As hitherto noted, exposing lower-level detail is more appropriately within the sphere of an LDM or PDM.

Familiar examples of entities, i.e., essential concepts predictably appearing in data models for conventional information systems, include the various roles assumed by persons (e.g., Employee, Manager, Salesperson, Student, Faculty Member, Contractor, Vendor Contact, etc.), places (Region, Country, State, Province, City, Street Address, etc.), business parties (e.g., Company, Government Agency, Customer, Supplier, Vendor, etc.), transactions between business parties (e.g., Request for Proposal, Purchase Order, Invoice, Receipt, etc.), and many other common intuitive entities such as organizations, departments, products, services, component parts, materials, events, assets, etc.

Note also that the term *entity* is frequently and equivalently referred to in data modeling literature as an *entity type* or an *entity class*. Software engineers and application architects, perhaps more accustomed to *object-oriented design* principles, are sometimes more comfortable with the term *object class* or simply *class*.[13]

Of course our immediate purpose is to focus on -- none other than -- the subject area of Amazon Database Services. Our intent is to provide a conceptual data model that concentrates on the essential entities which are fundamental to understanding each of the Amazon Database Services. You will ascertain, starting in Chapter 1, the truth that Amazon Database Service entities are much less concrete and more nebulous than the well-known previously identified examples.

A CDM's semantics also includes illustrating entity relationships which suggest associations, interactions, constraints, dependencies, assertions or rules between entities. Basically, relationships facilitate our overall understanding and reveal the manner by which entities are semantically connected with other entities. Examples of relationship assertions among entities such as "Supplier", "Purchase Order", and "Invoice" entities might be expressed as:

- "A Supplier can be issued one or more Purchase Orders."
- "A Purchase Order can be issued to at most one Supplier."
- "A Supplier can issue one or more Invoices."
- "Each Invoice requests payment for at least with one and possibly many Purchase Orders."

Also worth emphasis is the expectation that a conceptual data model provides pictures for illustrating entities and their relationships. Classically, data modelers have accomplished such by conceiving one or more entity-relationship diagrams. Such data model diagrams, if artfully and accurately crafted, often serve as a visual aid to abating ambiguity, enhancing communication between people, and demystifying complex concepts. The conceptual data model delivered herein will include many ERDs for vividly and graphically clarifying essential entities and relationships, thereby helping to illuminate the semantics and vocabulary of the Amazon database services.

Fortunately, a variety of data modeling tools, methodologies and graphical languages are available to data modeling practitioners to expedite the creation of ERDs. Conventions for depicting entities and relationships differ, depending on the specific tool or method used. ERDs routinely use boxes to depict entities. Lines annotated in various ways are normally used to depict relationships.

The DAMA-DMBOK cites *Information Engineering*,[11] promoted in books authored by James Martin, as the most common syntax and graphical approach to creating ERDs among data modeling practitioners.[14] We will apply Information Engineering diagramming conventions -- and specifically its "crow's feet" notation for depicting relationships -- as our preferred approach for creating ERDs. The tool selected to facilitate creating the diagrams herein is *ER/Studio Data Architect* (a product of IDERA, Inc.).[15]

Chapter 1 – Amazon EC2 Basics

Regions, Availability Zones, Images, Accounts

As noted in the introduction, Amazon database services are managed, platform services (PaaS) built on cloud infrastructure services (IaaS) provided by Amazon EC2. Understanding EC2 concepts is a prerequisite for understanding Amazon database services.

The entire infrastructure comprising Amazon EC2 (i.e., computer facilities, hardware, network, storage, and software resources) is geographically distributed among an ever growing number of globally dispersed and independent locations called *regions*. Basically, regions are established in separate geographic areas, typically in close proximity to major urban areas. Examples of names of Amazon EC2 regions contemporary to this publication include …

- US East (N. Virginia)
- US West (N. California)
- US West (Oregon)
- EU (Ireland)
- EU (Frankfurt)
- Asia Pacific (Tokyo)
- Asia Pacific (Seoul)
- Asia Pacific (Singapore)
- Asia Pacific (Sydney)
- South America (Sao Paulo)

Each region is made up of clusters of data centers known as *availability zones*. Inquisitive and astute readers are encouraged to reference the AWS Glossary to confirm the definition of a region -- which emphasizes the practicality of including a minimum of two availability zones within a region, and that a region, in effect, is a logical name for grouping the collection of all AWS resources within its locale.[16]

For availability zones, the AWS Glossary definition stresses that a region's availability zones are separate but interconnected physical environments, and are shielded from each other's local outages or infrastructure failures.

So, essentially, Amazon EC2 provides the ability to distribute cloud applications supported by virtual servers (aka virtual machines), network and data resources across multiple locations. Cloud application resources can be replicated across multiple regions for facilitating disaster recovery for an enterprise's mission-critical applications. Availability zones are isolated, independent, highly-available data centers and are generally unperturbed by failures occurring in other availability zones. By launching virtual servers in separate availability zones, applications are protected from failures in a single location, thereby enabling high availability.

Virtual servers in Amazon EC2 are simply called *instances*. The prerequisite starting point for launching an EC2 instance is called an Amazon Machine Image (AMI). Again, avid cloud students are invited to consult the AWS Glossary to corroborate the definition of an AMI, emphasizing that an AMI is a template (i.e., a master copy) the content of which is comparable to that of a server's root drive. As such, an AMI establishes the operating system and other basic software components (e.g., app server, web server or database server software) minimally necessary for standing up an instance. An AMI's content is encrypted and stored within AWS (e.g., using the Amazon Simple Storage Service -- Amazon S3).[17] So, an AMI essentially is a "launch pad" for propelling an instance into the AWS cloud. From an AMI (e.g., an AMI for initiating a Linux/Apache web server), many instances can be spawned which can be thought of as clones of the AMI and functioning as EC2 virtual servers.

AMIs and instances are examples of EC2 *resources* tied to an AWS *account*. The registration process for a new AWS customer results in creation of an AWS account for billing usage of AWS resources.

Again, voracious readers are encouraged to peruse the AWS Glossary to confirm the definition of an account. An account establishes information about the account owner including payment details for consuming any billable resources. An AWS account is also granted appropriate authorization for accessing and acting on any resources under its control. AWS account resources can be established in many regions. An account can own many AMIs.

Data Modeling Tutorial … Basics

Figure 1.1 graphically summarizes the previous narrative by unveiling our initial entity-relationship diagram. As previously noted, all entity-relationship diagrams herein illustrate Information Engineering notation conventions as implemented by IDERA's ER/Studio Data Architect. Figure 1.1 depicts fundamental entities and relationships for the EC2 concepts of region, availability zone, account, and image, and also facilitates a basic tutorial for introducing the aforementioned Information Engineering concepts and notation.

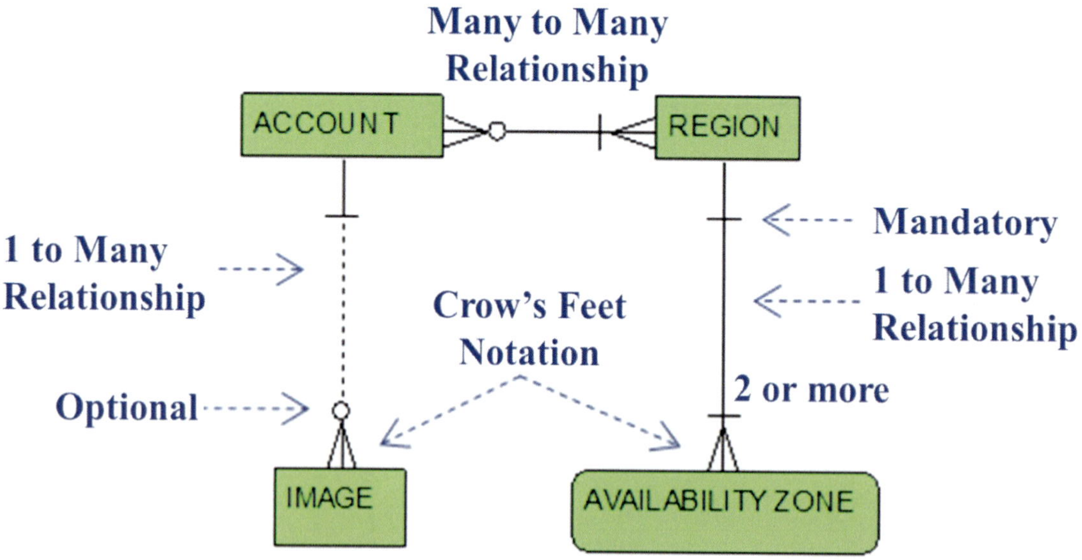

Figure 1.1 Regions, Availability Zones, Accounts, Images

The boxes in Figure1.1 represent entities. REGION, AVAILABILITY ZONE, IMAGE, and ACCOUNT are the names of entities for respectively maintaining information about regions, their associated availability zones, images accessible within a region, and accounts that own AWS resources, such as images. The REGION entity, for example, represents the set or collection of all regions within AWS. The AVAILABILITY ZONE entity represents the entire collection of all availability zones among all regions. The IMAGE entity represents the entire collection of all AMIs among all accounts. The ACCOUNT entity is the set of AWS accounts representing all AWS customers.

Solid or dashed lines connecting boxes represent *relationships* between those entities. Two *one-to-many* relationships and a single *many-to-many* relationship are shown in Figure 1.1. Note the trident-like "Crow's Feet Notation," intuitively indicating the "many" end of these one-to-many relationships and both ends of the many-to-many relationship.

From this diagram's graphical syntax, we can confidently infer that the one-to-many relationship between the REGION entity and the AVAILABILITY ZONE entity means that each region (i.e., each occurrence or member of the REGION entity, for example the "us-east-1" region) is made up of one or more availability zones (i.e., one or more members of the AVAILABILITY ZONE entity). More precisely, this relationship is annotated conforming to the AWS Glossary region definition, emphasizing that each region is minimally made up of two or more availability zones. Enabling high availability within a region makes no sense without at least two availability zones in the region.

Note the preference to use the terms "occurrence" or "member" of an entity, rather than the more commonly used data modeling terminology of an "instance" of an entity. The term "instance" of an entity is intentionally avoided herein so as not to be confused with an Amazon EC2 "instance."

Note also the short horizontal bar "-" crossing the relationship line at both ends of this relationship (i.e., near the box for the REGION entity and near the crow's feet for the AVAILABILITY ZONE entity). This denotes that the relationship is *mandatory* with respect to both entities. That is, each availability zone must be associated with one and only one region. In data modeling vernacular, the AVAILABILITY ZONE entity is referred to as a *dependent entity*. This means that each member of the AVAILABILITY ZONE entity cannot exist on its own and must be bound to a single region (i.e., a single member of the REGION entity). In other words, an occurrence of the AVAILABILITY ZONE entity exists only if it is inextricably linked to a single occurrence of the REGION entity. The relationship, of course, is also mandatory at the other end. By definition, a region requires at least two availability zones.

From Figure 1.1, we also infer that the one-to-many relationship between the ACCOUNT entity and the IMAGE entity means that each account (i.e., each occurrence or member of the ACCOUNT entity) can be the owner of one or more images (i.e., one or more members of the IMAGE entity).

Note also the circle "o" (denoting zero) perched on top of the crow's feet for this relationship. This symbolizes that the relationship is *optional* with respect to the IMAGE entity. This allows us to further infer that each occurrence of the ACCOUNT entity does not necessarily have any IMAGE members associated with it. More precisely, each account has zero, one or many associated images. Semantically equivalent, each account can have (or may have) one or more associated images, but might not have any images. When an account is initially created for a new user, there are no images associated with, i.e., owned by, the account. However, public images owned by other accounts can be selected within a newly created account.

Note also the short horizontal bar "-" crossing the relationship line at the opposite end of this relationship (near the box for the ACCOUNT entity). This denotes that the relationship is *mandatory* with respect to the ACCOUNT entity. That is, each image must be associated with one and only one account, i.e., the account that owns the image. The IMAGE entity is also a *dependent entity*, in that each occurrence of the IMAGE entity cannot exist on its own and must be bound to a single occurrence of the ACCOUNT entity. Each image exists only if it is joined to the hip of a single account.

In Figure 1.1, note the many-to-many relationship between the ACCOUNT entity and the REGION entity. This simply reflects that, when an account is established it is empowered to create and use EC2 resources in one or more regions and, clearly in the opposite direction, a region's resources can be accessed by many customer accounts.

Note also the difference between the dashed line relationship between ACCOUNT and IMAGE entities and the solid line relationship between the REGION and AVAILABILITY as annotated in Figure 1.2.

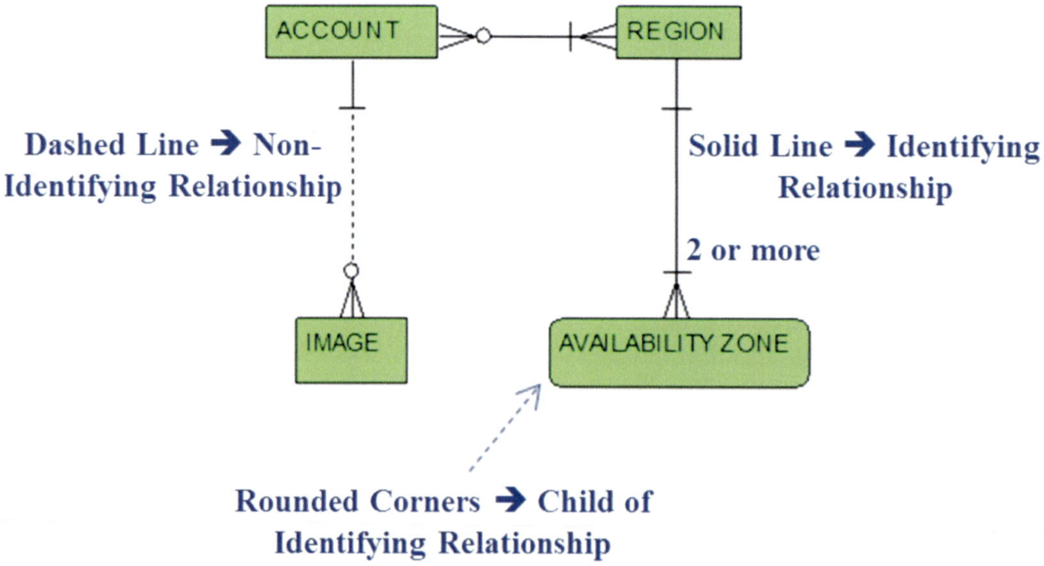

Figure 1.2 Identifying vs. Non-Identifying Relationships

A solid line for a one-to-many relationship denotes an *identifying relationship*. A dashed line denotes a *non-identifying relationship*. To appropriately clarify this difference, it is necessary to introduce the concepts of *attributes* and *foreign keys*, both additional fundamental data modeling concepts. Again, the DAMA-DMBOK comes to the rescue by imparting practical, no-nonsense definitions.[11]

First .... *Attributes* .... Simply stated, each entity is labelled typically by several (and sometimes by many) attributes. Each attribute is given a name and identifies some property, characteristic or information detail, the values for which facilitate an understanding of the entity. The values assigned to attributes are usually numbers, character strings, or dates. Typical examples of attributes in conventional information systems include *last name, first name, date of birth, address, salary, marital status, gender,* and *department* for example an EMPLOYEE entity.

Simple examples of attributes for the REGION entity are *region code* and *region name.* Values given to the *region code* attribute (e.g., the character strings "us-east-1" and "us-west-1") serve to uniquely identify each region, i.e., each member of the REGION entity.

The DAMA-DMBOK also divulges that there are two types of attributes: *key attributes or non-key attributes.*[11] Simply stated, a key attribute either by itself or grouped with other key attributes serves the purpose of uniquely differentiating a member (i.e., an occurrence) of an entity from any other entity member. By itself, the *region code* attribute of the REGION entity is sufficient to establish uniqueness among all REGION members and, therefore, serves as a *key attribute* (aka *primary key*).

Note that values of the *region name* attribute (e.g., "US East (N. Virginia)") are also unique. Therefore, *region name* could also be designated as a candidate key attribute. In such cases, one such attribute is designated as the key attribute, with remaining attributes relegated to the status of non-key attributes. It is also often the case that multiple key attributes as a group are collectively needed for determining uniqueness and are thereby jointly referred to as the primary key.

Next .... *Foreign Keys* .... In this case, the DAMA-DMBOK again admirably serves as a useful informant.[11] The concepts of foreign keys and key attributes are tightly coupled. In simplest terms, a foreign key consists of one or more attributes of an entity and creates a linkage to a related entity. More specifically, a foreign key provides a unique pointer to a member of the related entity.

For example, in a one-to-many relationship, the primary key (i.e., one or more key attributes comprising the primary key) migrates to and arrives as a foreign key in the entity on the "many" end of the one-to-many relationship. Key attributes of the primary key are thereby replicated (perhaps with different attribute names reflecting different roles) in the related entity. In a nutshell, a foreign key is an attribute or a cluster of attributes whose values match primary key values in a related entity, thereby serving as a technique for *joining* the related entities. A foreign key's attributes can also participate in the primary key of the entity to which they migrated.

Applying the foregoing explanation to the one-to-many *identifying relationship* between REGION and AVAILABILITY ZONE entities, the REGION entity's key attribute (i.e., *region code*) assists in identifying each occurrence within the AVAILABILITY ZONE entity. As actually implemented in EC2, the unique identifier for each availability zone consists of its associated *region code* and suffixed by a single character, e.g., "a", "b", "c", etc. For example, unique identifiers for availability zones within the "us-east-1" region (i.e., where region code = "us-east-1") include:

- us-east-1a
- us-east-1b
- us-east-1c
- us-east-1d
- us-east-1e

From the perspective of establishing a conceptual data model, however, an *AZ code* attribute can be defined for the AVAILABILITY ZONE entity (i.e., similar to *region code* for the REGION entity), whose values are single characters (as noted above "a", "b", "c", etc.). This *AZ code,* when concatenated as a suffix to the *region code,* serves to uniquely identify availability zones within an associated region. So, each occurrence of the AVAILABILITY ZONE entity is uniquely identified by the combination of both the *region code* and the *AZ code.*

From the previous foreign key explanation, the *region code* attribute of the REGION entity migrates to the AVAILABILITY ZONE entity as a *foreign key* attribute. Thus, in this sense, the *region code* attribute for REGION occurs in both entities, thereby joining both entities. The full identifier for each availability zone is the combination of the value held in the *region code* attribute (i.e., the foreign key that migrated from the REGION entity) in combination with the value held in the *AZ code* attribute of the AVAILABILITY ZONE entity.

Therefore, from the perspective of a conceptual data model, one can argue that the relationship between REGION and AVAILABILITY ZONE entities is an identifying relationship. In reality however (i.e., from the standpoint of a physical database design), these are inseparably combined into a single attribute which, in EC2 API documentation, is referred to as the *availability zone id.* Nonetheless, *region code* is an indelible component of an availability zone's identity, illustrated as follows.

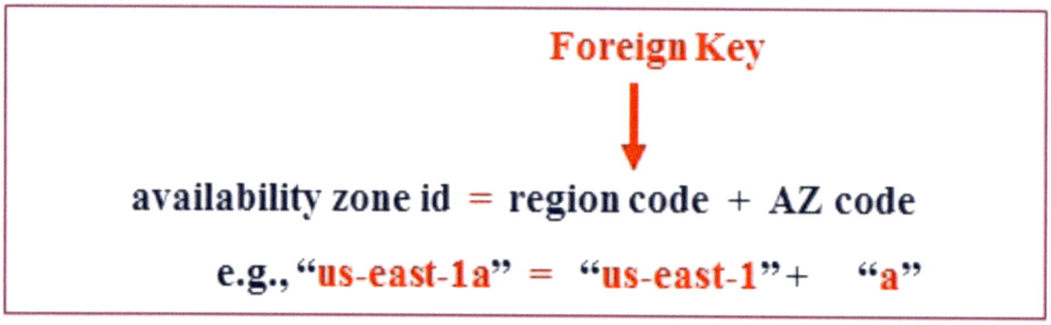

Note also the rounded corners for the depiction of the AVAILABILITY ZONE entity in both Figures 1.1 and 1.2. ER/Studio distinguishes dependent entities participating as the child entity of an identifying relationship (i.e., the entity at the "many" end of an identifying relationship) with rounded corners.

A final comment is worthy of emphasis regarding properly setting expectations for the scope of a conceptual data model relative to specifying attributes of entities. Drawing from Andy Graham's book[18] entitled The Enterprise Data Model: A framework for enterprise data architecture ...

> *Conceptual data models ... are used to understand the high level data entities for the domain in question ... Conceptual models are not like your normal data model as they don't contain much in the way of rigor. What we are after is the general feel for the key data entities and relationships. This model forms the framework we can use to undertake a more detailed analysis and design.*

To summarize, a conceptual data model's purpose is to provide a high-level, visually intuitive description of potentially complex concepts and their relationships *for the domain in question*, which of course in this case is Amazon EC2. As a result, the intent is not to provide exhaustive detail in terms of specifying all attributes belonging to each entity. Such is the objective of a more rigorous logical data model (LDM), which exploits the conceptual data model as its starting point. Attributes will be introduced only as necessary and appropriate, i.e., where it makes sense to enrich our understanding of Amazon EC2's essential entities and relationships.

Instances and Instance Types

As introduced in the previous section, EC2 instances are simply virtual servers launched from AMIs. In the context of EC2, the term *instance* is really just a shortcut for *instance of a virtual server*. Once again, perspicacious readers are invited to consult the AWS Glossary to inspect the simplicity of the instance definition -- emphasizing that an *instance* is reproduced as a copy of an AMI and is spawned as a virtual server in the AWS cloud.

A single image can be used to launch many instances. An instance cannot be created without referencing a single image. After an instance has been launched, its configuration of installed software components can be modified. For example, an instance launched from a basic LINUX image can then later be provisioned with Apache, thereby resulting in a web server instance. A new AMI can then be created from this modified instance.

When launching each instance, in addition to specifying an image, an *instance type* must also be specified indicating the sizes of the instance in terms of its number of virtual CPUs and the amount of memory and storage. The AWS Glossary further emphasizes that different instance types are intended for different purposes depending on the performance requirements of an application. Some instance types are more appropriate for standard applications characterized with low or moderate load requirements, whereas other more potent and higher-priced instance types are suitable for applications with high throughput or rapid response time requirements.

Each instance type offers different compute, memory and storage capacities. For example, names of "General Purpose" instance types include a1.medium, a1.large and a1.xlarge. Examples of "Compute Optimized" instance types include c5.large, c5.xlarge and c5.2xlarge. "Memory Optimized" instance types include r5.large, r5.xlarge and r5.2xlarge.

Figure 1.3 extends Figure 1.1 by including instance and instance type related entities and relationships.

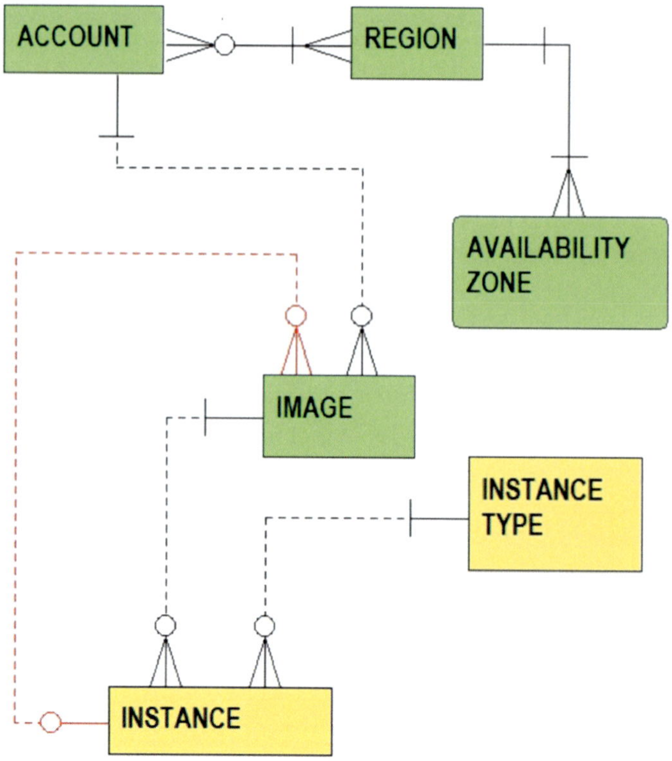

Figure 1.3 Instances and Instance Types

From Figure 1.3, the one-to-many relationship from the INSTANCE TYPE entity to the INSTANCE entity reflects that each instance must be associated with a single instance type. Each instance type can apply to many instances. The one-to-many relationship from the IMAGE entity to the INSTANCE entity expresses that an instance must be generated, i.e., it must have been created, from a single image, and that an image can serve as the template for many instances. Note also the return-trip one-to-many relationship from the INSTANCE entity back above to the IMAGE entity. This emphasizes that an instance launched from an image can then be modified by installing additional software components. A new image can then be created from this modified instance. In the fullness of time, additional modifications are made to an instance, and it is possible that many different images can be spawned from the same instance.

VPCs, Subnets, IP Address Ranges, Security Groups

Amazon EC2 enables each account to establish one or more *Virtual Private Clouds* (VPCs). In simplest terms, a VPC is a virtual network functionally analogous to a conventional physical network typical of any corporate data center. Once again, unremitting and persevering cloud students might be motivated to consult the AWS Glossary to enjoy the simplicity of the VPC definition -- emphasizing flexibility of access by VPC resources to AWS services, shared connectivity and security facilities.

Essentially, a VPC is a network construct for segregating virtual servers and each VPC is logically separate from other VPCs in the AWS cloud. AWS resources such as EC2 instances, virtual firewalls, and other networking and security resources can be launched into a VPC. A range of IP addresses is apportioned to each VPC. Instances launched within a VPC are then provisioned with one or more IP addresses within this assigned IP address range.

For example, a corporate enterprise can establish several VPCs for distinct reasons, such as for different organizations, product lines, profit centers, etc. Different VPCs can also be set up for various operational needs, such as establishing separate VPCs quarantining resources for applications development, quality assurance and production applications.

Each VPC can traverse more than one availability zone within the same region. Each VPC is composed of one or more *subnets,* each with its own subset of IP addresses bounded within the same IP address range assigned to the VPC. Each subnet must be associated with a single VPC and a single availability zone. A second subnet within a VPC can be associated with a different availability zone. For high availability, a cloud application, therefore, can be established with instances in one subnet of a VPC and failover resources in a second subnet of the same VPC. But the second subnet is in a different availability zone.

The *subnet* definition from the AWS Glossary also emphasizes that an instance is assigned IP addresses within its subnet's IP address range and that AWS resources can be launched and grouped into a selected subnet based on security requirements or operational needs.

Scenarios for establishing subnets satisfying special security and operational requirements include …

- Use of a *public subnet* (aka a DMZ) for Internet facing resources, i.e., accessible from the Internet, such as load balancers or web servers responding to external requests from an enterprise's customers.
- Use of a *private subnet* for resources that are not directly accessible from the Internet, e.g., for web servers, application servers, and database servers that need to be accessed by resources launched within a public subnet.
- Use of a *private subnet* for resources that are reserved for an enterprise's internal applications.
- Subnets can also be established for segregating resources for applications development, QA, and production.

Figure 1.4 extends Figure 1.3 by including VPC and subnet related entities and relationships.

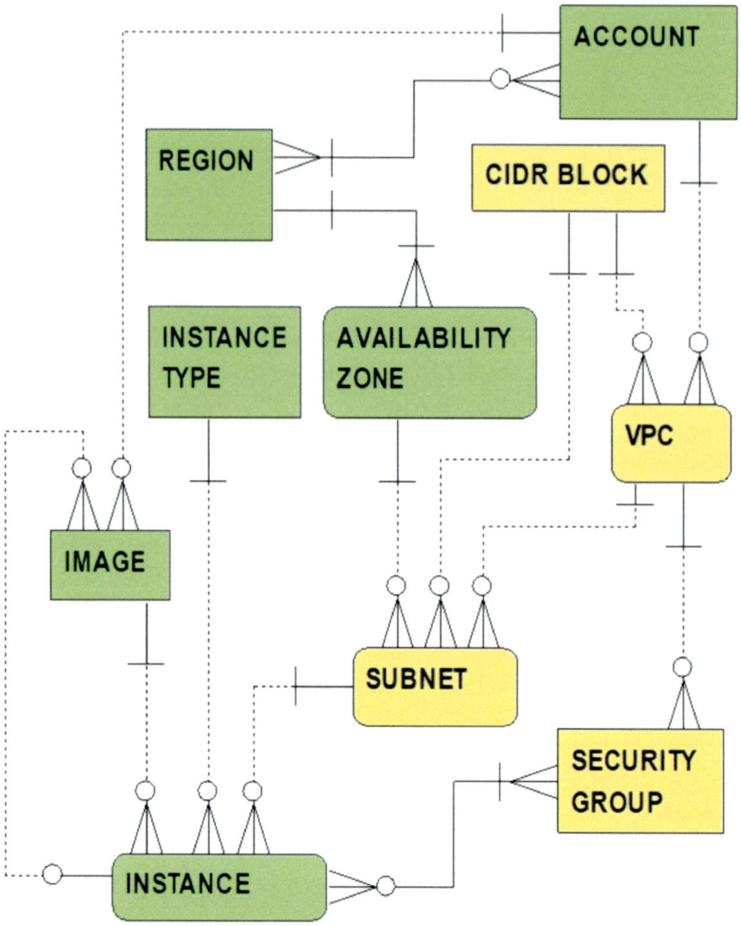

Figure 1.4 VPCs, Subnets, IP Address Range = CIDR Block, Security Groups

From Figure 1.4, the one-to-many relationship from the ACCOUNT entity to the VPC entity reflects that each VPC must be associated with a single account. In the other direction, each account can own many VPCs. The one-to-many relationships from the VPC and AVAILABILITY ZONE entities to the SUBNET entity expresses that each subnet must be tied to a single VPC and a single availability zone. In the opposite direction, an availability zone can host many subnets and a VPC can be composed of multiple subnets. The SUBNET entity, in effect, resolves an implicit many-to-many relationship between the AVAILABILITY ZONE entity and VPC entity. Accordingly, a VPC can traverse multiple availability zones, and each availability zone can host multiple VPCs, by virtue of the availability zone hosting VPC subnets. Note also the one-to-many relationship from the SUBNET entity to the INSTANCE entity. This emphasizes that each instance launched within a VPC must be tied to a single subnet. Each subnet can be associated with many instances.

Also in Figure 1.4, note the CIDR BLOCK entity. An IP address range is also known as a *CIDR block*.[19] A glimpse at the AWS Glossary's CIDR block definition reveals that *CIDR* is an abbreviation for *Classless Inter-Domain Routing,* which is a structured method for allocating IP addresses and for facilitating the routing of IP packets between and within IP networks.

The one-to-many relationships from the CIDR BLOCK entity to the VPC and SUBNET entities express that each VPC must be tied to a single CIDR block and each subnet must also be associated with a single CIDR block. A subnet's CIDR block is an IP address range which is a subset within the boundary of its parent VPC's IP address range.

The IP address ranges represented by these CIDR blocks consist of a series of private IP addresses, which are not publicly accessible via the Internet. These IP addresses are also private in the sense that they are reserved for enabling interaction between instances within the same VPC. These CIDR blocks are specified using what is known as Internet Protocol version 4 (IPv4) notation. It is possible to assign the same IPv4 CIDR block to multiple VPCs within the same account or among different accounts.

Also in Figure 1.4, note the SECURITY GROUP entity. Amazon EC2 is endowed with virtual firewalls referred to as *security groups* for governing and securing traffic to and from EC2 instances. Once again, astute readers are encouraged to seek guidance from AWS Glossary's *security group* definition. This definition draws attention to the many-to-many relationship between security groups and instances. More specifically, a security group can act as a firewall for many instances, and traffic for each instance can be regulated by more than one security group. At launch time, an instance can be coupled with one or more security groups. For each security group, firewall rules are specified for allowing inbound traffic or outbound traffic, valid source and target IP address ranges, and permitted communication protocols (e.g., TCP, UDP, and ICMP). The one-to-many relationship between the VPC and SECURITY GROUP entities exhibits that each VPC can be associated with many security groups. The many-to-many relationship between the INSTANCE and SECURITY GROUP entities indicates that each instance must be fire-walled by at least one security group but possibly can belong to many security groups, and that each security group can govern traffic for many instances.

Instances ... VPC vs. Classic

There are two general categories of EC2 instances: *classic instances* and *VPC instances*. At the dawn of AWS (i.e., for the initial releases of the AWS cloud) the AWS world was flat. The concept of VPCs had not yet arrived. The AWS cloud conceptually was a single level, horizontal network cloud that was common and shared by all AWS customers. VPCs were introduced to AWS late in 2013.[20] EC2 instances prior to the introduction of VPCs are known as classic instances. Legacy support is provided for classic instances of elderly AWS accounts that were created prior to the unveiling of VPCs. Newer accounts support only VPC instances, i.e., instances supported under the umbrella of a VPC.

Classic instances, though still supported, are a remnant of antiquity, not available and disavowed for new AWS accounts. A classic instance is directly associated with its legacy account. Each VPC Instance is indirectly associated with its account via its subnet and VPC. Figure 1.5 extends Figure 1.4 by adding *sub-types* for the INSTANCE entity.

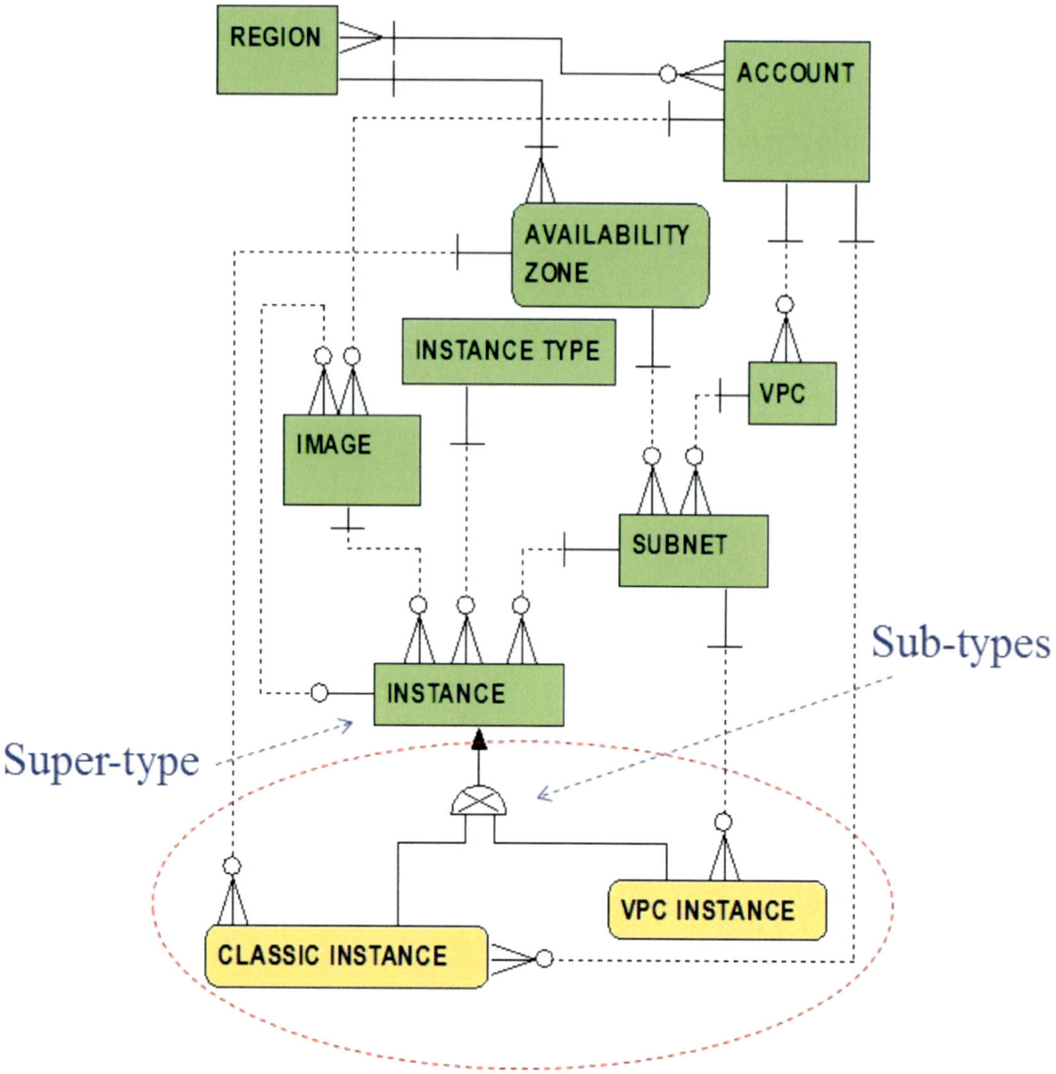

Figure 1.5 Instance Sub-types: Classic vs. VPC Instances

From Figure 1.5, in data modeling vernacular, the INSTANCE entity is referred to as a *super-type* of both the CLASSIC INSTANCE and VPC INSTANCE entities. CLASSIC INSTANCE and VPC INSTANCE entities are referred to as *sub-types* of the INSTANCE super-type. In this case, the diagram (via the semi-circle with the embedded "X") conveys that an instance must be either a classic instance or VPC instance; or, more precisely, a member of the INSTANCE entity must be either a member of the CLASSIC INSTANCE entity or a member of the VPC INSTANCE entity.

In many data modeling references, a description of super-types and sub-types is coupled with the concept of *generalization*. Again appropriating from the DAMA-DMBOK, one ascertains that a super-type is a generalization of its sub-types.[11] This means that a super-type is defined with attributes and relationships which are generic to (i.e., shared by) all of its sub-types. Each sub-type *inherits* such common attributes and relationships from its super-type but also differentiates itself with additional attributes or relationships peculiar to the sub-type. Each sub-type is in effect a special case of the super-type.

So the INSTANCE entity is a generalization of the CLASSIC INSTANCE and VPC INSTANCE entities. A CLASSIC INSTANCE is *a kind of* INSTANCE and a VPC

INSTANCE is *a kind of* INSTANCE. CLASSIC INSTANCE and VPC INSTANCE entities are special cases of the INSTANCE entity. An instance is exclusively either a classic instance or a VPC instance but not both. Note the italicized phrase "*a kind of*" is commonly used in data modeling literature to describe an inheritance relationship in the direction of a sub-type to its super-type (i.e., a sub-type is "*a kind of*" its super-type).[11] This phrase will continue to be italicized herein to subliminally emphasize such customary usage.

Rules and relationships are different depending on the sub-type. Classic instances are differentiated from VPC instances to show special relationships and rules relative to other Amazon EC2 concepts. In this case, Figure 1.5 differentiates the direct relationship from the CLASSIC INSTANCE sub-type to the ACCOUNT entity, the direct relationship from the VPC INSTANCE sub-type to the SUBNET entity, and allows one to infer an indirect relationship from the VPC INSTANCE sub-type to the ACCOUNT entity (via SUBNET and VPC entities).

But again, it is worth emphasizing that classic instances can be considered an embryonic concept, grandfathered from the past, applying only to legacy AWS accounts. In the spirit of simplification, therefore, the conceptual data model evolving herein will adopt a viewpoint demoting classic instances from further consideration. Henceforth, it is assumed in this conceptual data model that all instances are VPC instances. The shunning of classic instances is illustrated as follows in Figure 1.6.

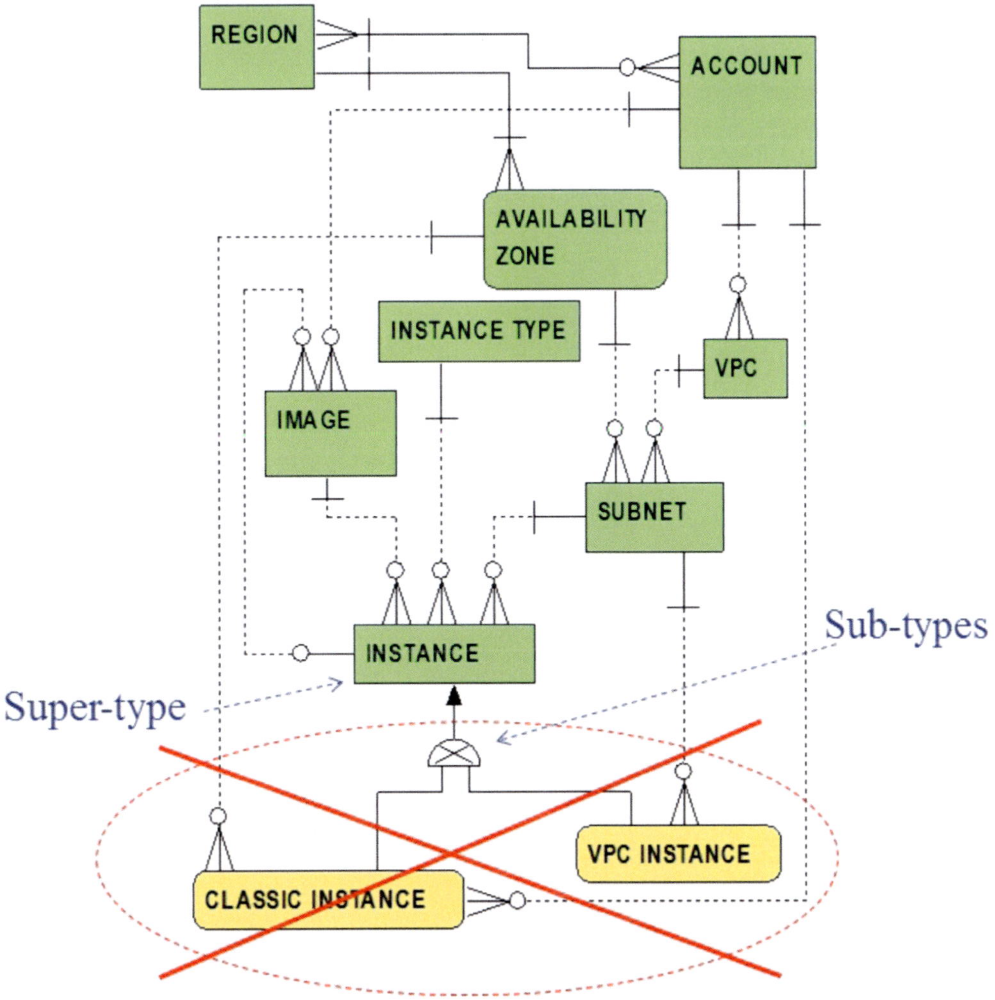

Figure 1.6 Classic Instances Not In Scope

Figure 1.6 highlights that classic instances are removed and exempted from the scope of this conceptual data model. Instances hereafter are declared to be VPC instances, thus substantially simplifying the data model. The INSTANCE entity reflects VPC instances only.

Resources, Tags

Amazon uses the term *resource* as a generalization for many of the basic EC2 concepts reflected as fundamental entities in the previously portrayed ERDs. Resource examples include VPCs, instances, images, security groups, and subnets. Essentially, a *resource*, as informally inferred from its AWS Glossary definition, is any AWS entity that a customer is able to act on, control or operate (e.g., by exercising the Amazon EC2 console, API, or CLI). A unique resource identifier -- referred to as an Amazon Resource Name or ARN -- is assigned to each resource when it is created.

Resources can also be anointed with user defined attributes known as *tags*. In basic terms, again inferring from its AWS Glossary definition, a tag is simply customer defined metadata, like a label attached to a resource. Tags can be declared to flexibly assign any additional customer-specific information, for example to further categorize or identify resources. Each tag consists of a key (i.e., a unique string of characters), and an optional value, which is also a string of characters.

Examples of resource tags might include an *account number* and *department* that will be billed for the resource, the *owner* of the resource, a brief description of its *purpose*, etc. The semantics of each tag is entirely within the customer's purview and interpretation.

Figure 1.7 depicts essential entities supporting resources and tags.

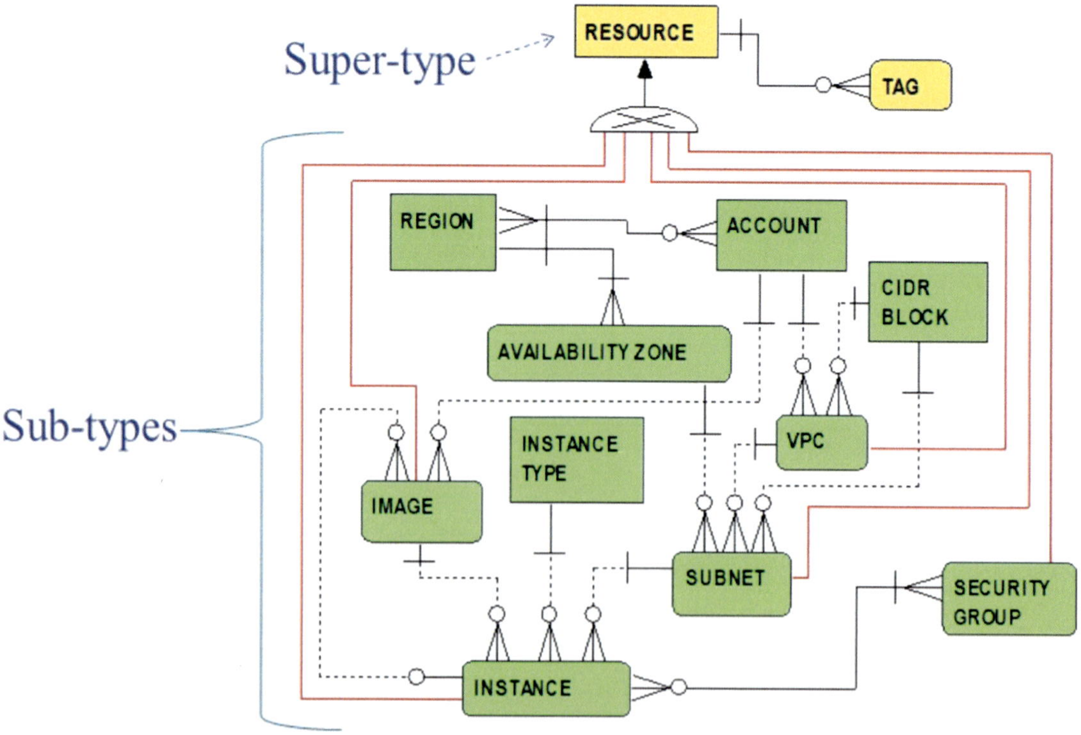

Figure 1.7 Resources and Tags

From Figure 1.7, The RESOURCE entity is shown as a super-type having many resource sub-types, each corresponding to previously introduced EC2 essential entities. VPC, SUBNET, IMAGE, INSTANCE, SECURITY GROUP sub-types are a subset of all EC2 resources. Many additional entities will be introduced in subsequent chapters which will be identified as resource sub-types. Each sub-type of the RESOURCE entity inherits the ARN attribute as a unique identifier. The one-to-many relationship between the RESOURCE and TAG entities reflects that each resource can be assigned one or more customer defined tags. This tagging relationship, of course, is inherited by all resource sub-types. Concurrent with this publication, tags are not enabled for all AWS resources. Tagging is supported for each of the sub-types illustrated in Figure 1.7.

Chapter Summary

Figure 1.8 summarizes essential entities and relationships for Amazon EC2 basic concepts.

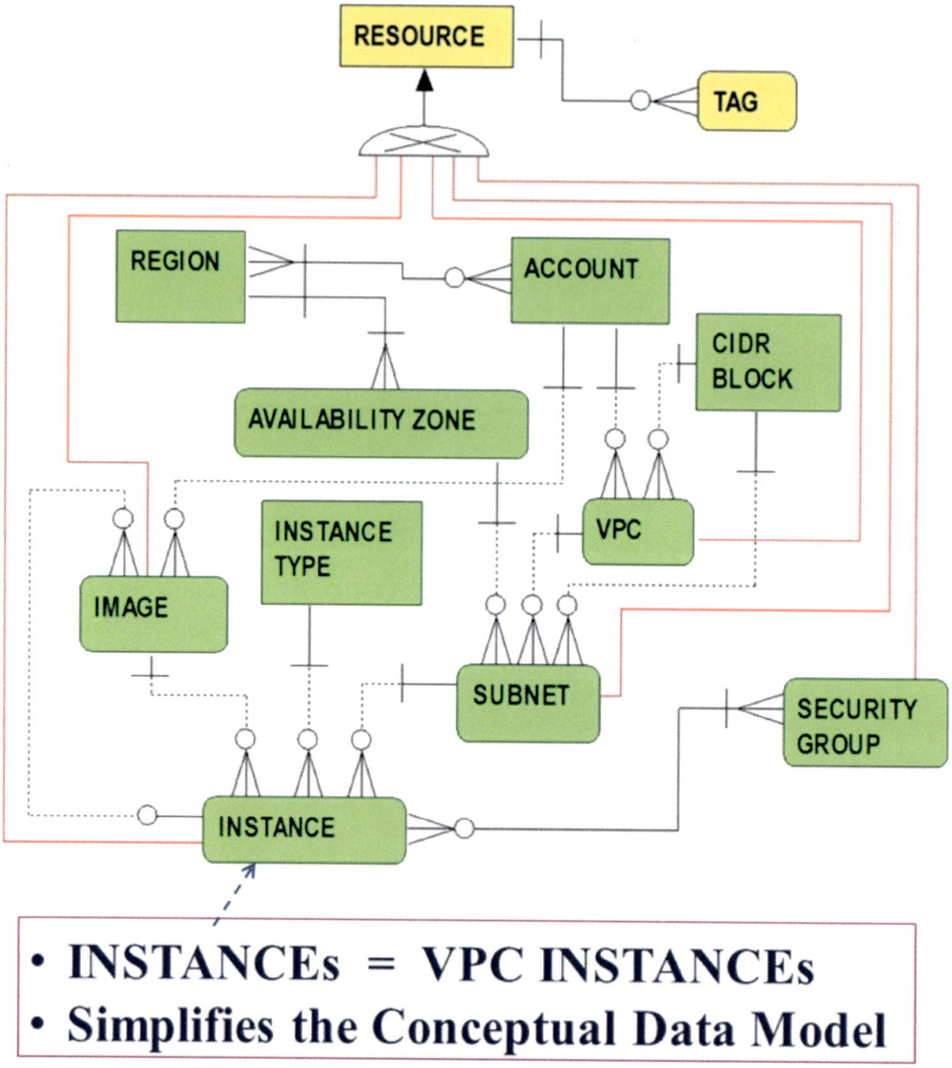

Figure 1.8 Amazon EC2 Basics Summary

Data modeling ERDs illustrate Information Engineering notation and depict fundamental entities and relationships for EC2 concepts. The boxes in ERDs represent entities. Solid or dashed lines with "Crow's Feet Notation" connecting boxes represent relationships such as one-to-many relationships and many-to-many relationships. An attribute is a property of an entity whose values help describe an occurrence of an entity. Attributes assigned to entities are either *key* or *non-key attributes.*

A key attribute either by itself or grouped with other key attributes uniquely identifies a member of an entity and is referred to as a *primary key*. A primary key can become a *foreign key* attribute of another related entity. Relationships can be mandatory or optional. A solid line for a one-to-many relationship denotes an *identifying relationship,* where a foreign key contributes to its entity's primary key. A dashed line denotes a *non-identifying relationship,* where a foreign key does not participate in its entity's primary key.

Amazon EC2 infrastructure is distributed among globally dispersed geographic areas, called *regions,* established close to major urban areas, e.g., the US East (N. Virginia) region. Each region is made up of clusters of nearby data centers known as *availability zones.* Amazon EC2 enables distribution of cloud applications supported by virtual servers (called EC2 *instances*), network and data resources replicated across multiple regions for ensuring disaster recovery. Availability zones are isolated, independent and enable high availability. By launching instances in separate availability zones, applications are protected from failures in a single location.

The starting point for an EC2 instance is called an Amazon Machine Image (AMI), which encapsulates the software components minimally necessary for launching an instance in EC2, e.g., an AMI for a Linux server with Apache installed. Many instances can be spawned from an AMI. AMIs and instances are examples of EC2 *resources* tied to an AWS *account* for establishing ownership and billing AWS resource usage. AWS account resources can be established in many regions. An account can own many AMIs. After an instance has been launched, it's configuration of installed software components can be modified. A new image can be fashioned from this modified instance. When launching each instance, in addition to specifying an image, an *instance type* must also be specified for indicating the size of the instance in terms of virtual CPUs, memory and storage.

Each account can be endowed with one or more Virtual Private Clouds (VPCs). A VPC is a virtual network functionally akin to a physical network typical of corporate data center networks. Each VPC is logically separate from other VPCs. AWS resources such as instances, virtual firewalls, and other networking resources can be launched into a VPC. A CIDR block consisting of a range of IP addresses (IPv4) is apportioned to each VPC for provisioning private IP addresses to instances launched within the VPC.

Separate VPCs can be established for different organizations, product lines, profit centers, etc. A VPC can also be established for various operational needs, such as for applications development, QA, and production applications. Each VPC is composed of one or more *subnets,* each allotted its own subset of private IP addresses within the same CIDR block (IPv4) allocated to the subnet's VPC.

Each subnet must be associated with a single VPC and a single availability zone. A second subnet within a VPC can be associated with a different availability zone for high availability. Subnets can be established based on security or operational needs, such as *public subnets* for supporting Internet facing applications, and *private subnets* for serving an enterprise's internal applications or for segregating resources for applications development, QA, and production.

Amazon EC2 has virtual firewalls referred to as *security groups* for governing and securing traffic to and from EC2 instances. A security group can act as a firewall for many instances, and traffic for each instance can be regulated by more than one security group. Firewall rules are specified for allowing inbound or outbound traffic, source and target IP address ranges, and permitted communication protocols (e.g., TCP, UDP, and ICMP). Each VPC can have many security groups. Each instance must be fire-walled by at least one security group but possibly belongs to many security groups. Each security group can govern traffic for many instances.

There are two general categories of EC2 instances: *classic instances* and *VPC instances*. EC2 instances created via early AWS releases and prior to the introduction of VPCs are known as classic instances and continue to be supported for long-standing accounts. Newer accounts support only VPC instances. For simplicity sake, the scope of this data modeling effort is focused on VPC instances.

Amazon uses the term *resource* as a generalization for many of the basic EC2 concepts. Resource examples include VPCs, instances, images, security groups, and subnets. Essentially, a *resource* is any AWS entity that a customer is able to act on, control or operate. A unique resource identifier, an ARN, is assigned to each resource when it is created. Resources are labelled with user defined attributes known as *tags* -- customer defined metadata. Tags can be declared to flexibly assign any additional customer-specific information.

Chapter 2 – Amazon Relational Database Service (RDS)

Amazon RDS Introduction

EC2 serves as the foundation for scores of additional Amazon web services. Very prominent among these services is the Amazon Relational Database Service (Amazon RDS), a *managed* web service for enabling and facilitating the implementation of familiar, contemporary relational database management systems (RDBMS) -- such as Oracle DB, Microsoft SQL Server, MySQL, MariaDB, and PostgreSQL -- in the Amazon cloud.

As a managed service, RDS facilitates and reduces the stress of dealing with administrative chores for overseeing the operational aspects of relational databases. Mundane admin tasks usually performed by DBAs (e.g., scaling additional CPU resources, allocating additional memory and storage devices, performing backups, applying software patches, detecting and reacting to database failures, recovering databases from backups, etc.) are simplified and often expedited by RDS.

RDS also accelerates the implementation of complex requirements, such as *high availability* by optionally synchronizing database updates and failing over to a secondary database instance in the event of database failure. In some cases (i.e., for MySQL, MariaDB, and PostgreSQL), RDS enables the establishment of *read replicas* -- additional database instances for read-only access -- to relieve the pressure on a primary database instance from query activity.

Figure 2.1 introduces RDS basic entities and relationships.

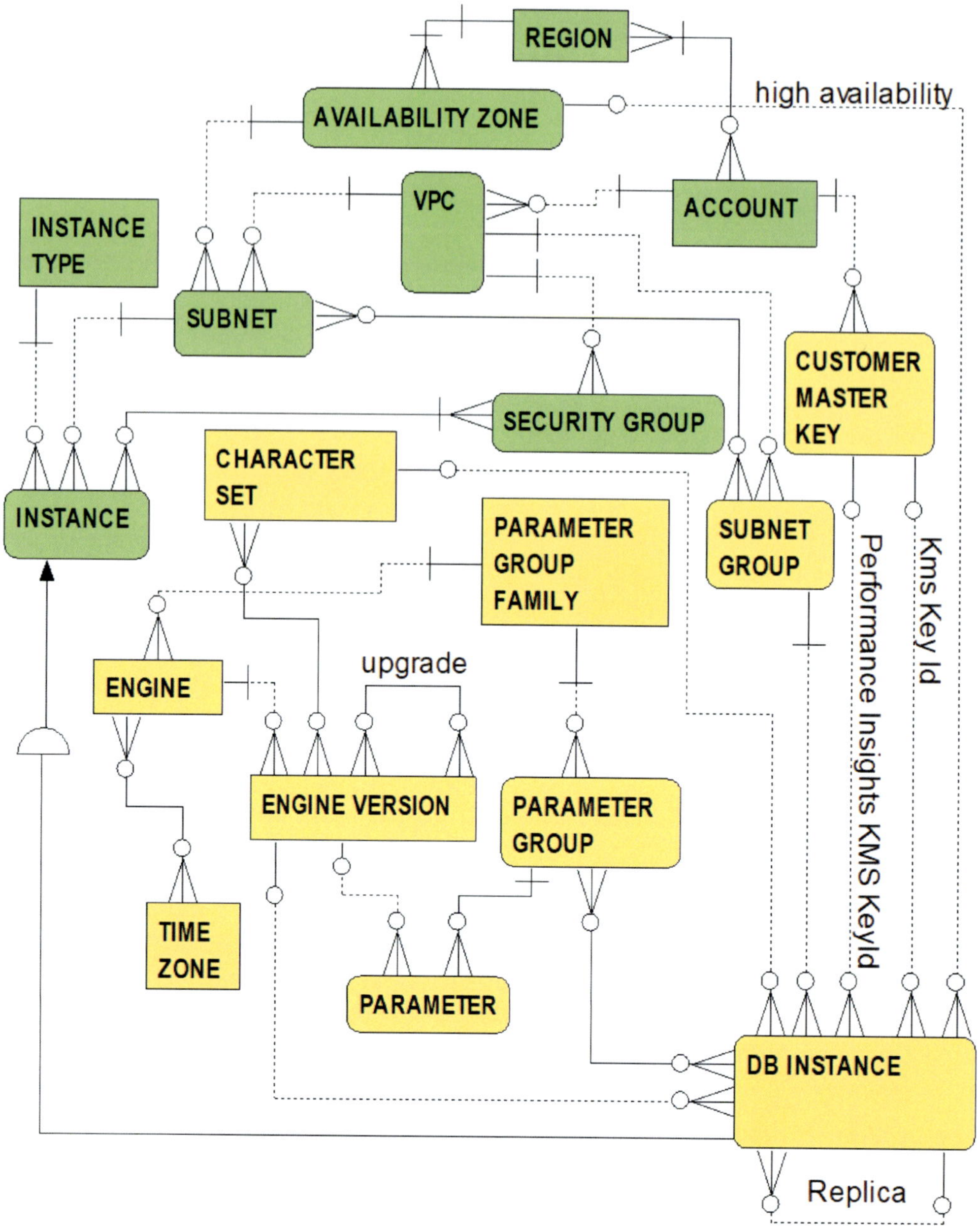

Figure 2.1 RDS Basic Entities and Relationships

In Figure 2.1, the ENGINE and ENGINE VERSION entities reflect the DBMS engines and versions of database engines supported by RDS (e.g., the various supported versions of Oracle DB, MySQL, PostgreSQL, Microsoft SQL Server, and MariaDB). Each DBMS engine has many supported versions. The many-to-many relationship between the CHARACTER SET and ENGINE VERSION entities emphasizes that each version of a database engine can support many character sets, one of which can be designated as the default character set. In the opposite direction, a character set (e.g., UTF8) can be supported by many engine versions. The many-to-many reflexive relationship (labelled "upgrade") between the ENGINE VERSION entity and itself points out possible engine versions that are valid targets for upgrade. More specifically, each engine version can possibly be upgraded to many later and greater versions, and an engine version can be the target of many prior engine versions.

The PARAMETER GROUP and PARAMETER entities deal with maintaining database configuration information. The one-to-many relationship between the PARAMETER GROUP and PARAMETER entities conveys that each parameter group is a collection of many parameters used for setting up a database. Values are assigned to database parameters, such as for the database time zone, character set, optimization features, audit trails, cache size for caching tables, the maximum number of processes for parallel execution, various categories of database memory, etc. More specific examples, well-known to Oracle DBAs, include setting parameter limits for *system global area* (i.e., SGA, the memory shared among processes within an Oracle database), and *program global area* (i.e., PGA, the private memory reserved for a single Oracle database process). Names of specific SGA and PGA related parameters for which values can be set include *sga_target*, *sga_max_size*, and *pga_aggregate_target*. The following example of an RDS CLI command, *describe-engine-default-parameters,* can be used for providing default parameters for a specific engine family (i.e., in this case for "oracle-ee-12.1").

aws rds describe-engine-default-parameters
--db-parameter-group-family oracle-ee-12.1

For setting the time zone parameter of a database, note the many-to-many relationship between the ENGINE and TIME ZONE entities emphasizing that each database engine supports many time zones and, of course, a time zone can be supported by many engines. For example, the default time zone for MySQL and MariaDB databases is *Universal Time Coordinated* (UTC). For these engines, however, the *time_zone* parameter in a parameter group can be set to the specific regional time zone (i.e., set to one of the supported values associated with the database engine) corresponding to the application's time zone.

The one-to-many relationship between the ENGINE VERSION and PARAMETER entities conveys that each parameter can be associated with an initial engine version (i.e., the initial engine version for which the parameter is relevant).

The DB INSTANCE entity represents a collection of EC2 instances for running isolated database environments. Such is emphasized in Figure 2.1 by noting that DB INSTANCE is a sub-type of the INSTANCE entity, thereby inheriting attributes and relationships of its INSTANCE super-type (e.g., each DB instance inherits an instance type, resides in a subnet, is associated with security groups, etc.) because a DB instance is *a kind of* EC2 instance. Depending on the database engine, each DB instance can support a single database (e.g., an Oracle database with multiple *schemas*) or multiple databases (e.g., a Microsoft SQL Server environment).

The one-to-many relationship between the ENGINE VERSION and DB INSTANCE entities emphasizes that each DB instance is associated with a single DBMS engine version. The one-to-many relationship between the CHARACTER SET and DB INSTANCE entities emphasizes that each DB instance is associated with a single character set. The one-to-many relationship between the AVAILABILITY ZONE and DB INSTANCE entities emphasizes that each DB instance can be associated with an alternate availability zone for implementing high availability. As previously noted, RDS provides an option for having a secondary DB instance synchronized with the primary DB instance for failover to the secondary DB instance residing in this alternate availability zone.

The one-to-many relationships (i.e., two relationships) between the CUSTOMER MASTER KEY and DB INSTANCE entities emphasize two database encryption options available to each DB instance. Amazon encryption is provided by yet another service, the AWS Key Management Service (KMS), which uses *customer master keys* (CMKs) as the basis for encrypting or decrypting data.[21] The two relationships between CUSTOMER MASTER KEY and DB INSTANCE entities imply that two CMKs are optionally provided as foreign keys (FKs) to a DB instance. The relationship labelled "KMS Key Id" is the FK used for encrypting DB instance data. Amazon RDS is also endowed with the ability to monitor the load and analyze the performance of DB instances. RDS refers to this as *Performance Insights data* invaluable for troubleshooting database performance problems. The relationship labelled "Performance Insights KMS Key Id" is the FK for encrypting such performance metrics. This capability is supported for only a subset of RDS database engines and versions.

The context and purpose of the SUBNET GROUP entity is clarified with the help of the many-to-many relationship between the SUBNET GROUP and SUBNET entities. A *subnet group* is simply a collection of subnets and each subnet can belong to many subnet groups. The one-to-many relationship from the VPC entity to the SUBNET GROUP entity suggests that each subnet group is defined within a VPC. The one-to-many relationship between the SUBNET GROUP and DB INSTANCE entities suggests that each DB instance is associated with a single subnet group, which is used by RDS for assigning the DB instance to an appropriate subnet. Each subnet group should have subnets dispersed among multiple availability zones to facilitate implementing high availability (i.e., RDS can use an alternate subnet in a different availability zone if the primary DB instance fails).

The many-to-many relationship between the PARAMETER GROUP and DB INSTANCE entities shows that a DB instance can actually be associated with more than one parameter group. This accounts for the fact that RDS supports replacing the currently assigned parameter group with a second parameter group and that the status of this modification is *pending*. Most commonly, a DB instance has only one parameter group associated with it.

The one-to-many reflexive relationship between the DB INSTANCE entity and itself emphasizes RDS support, as previously noted, for *read replicas* applicable to MariaDB, MySQL, and PostgreSQL database engines. A primary DB instance serves as the source of data copied to one or more DB instances serving as read replicas. A read replica provides read-only access. Rather than accessing the primary DB instance, an application can re-direct its database queries to a read replica, thereby alleviating the workload on the primary DB instance. Any data updates on the primary DB instance are automatically and asynchronously pushed to its read replicas. Read replicas can be added as-needed for scaling capacity based on actual utilization.

In addition to attributes derived from the above noted one-to-many relationships to the DB INSTANCE entity (i.e., foreign keys), noteworthy examples of additional attributes of the DB INSTANCE entity include:

- *DB instance identifier*: A unique identifier provided by the user for the DB instance.

- *DB instance status*: Indicates the current DB instance status or health (e.g., "available", "backing-up", "creating", "deleting", etc.).
- *storage encrypted*: Indicates whether or not DB instance data is encrypted.
- *backup retention period*: The number of days that automatic backups are saved prior to being purged.
- *multi AZ*: Indicates if multiple availability zones for high availability applies to this DB instance (i.e., using data replication to a fall-back DB instance in another availability zone).
- *preferred backup window*: If automatic backups are enabled, this indicates the time range (daily) when automated backups can be created.

Option Groups and Options

Database engines provide optional capabilities, features or tools that extend or enhance fundamental engine capabilities. Amazon RDS refers to such features as *options* which are collected into *option groups*. Option groups are associated with DB instances. Examples of options for Oracle include:

- Oracle Enterprise Manager (OEM)
- Oracle Java Virtual Machine (JVM)
- Oracle Native Network Encryption (NATIVE NETWORK ENCRYPTION)
- Oracle Statspack (STATSPACK)
- Oracle Time Zone (TIMEZONE)
- Oracle XML DB (XMLDB)

Options have a collection of *option settings* that are specified to enable, configure, or govern the behavior of these capabilities for a DB instance. Each option setting has a current value and can also specify the set of possible valid values. For example, the Oracle Native Network Encryption option has an option setting "SQLNET. ENCRYPTION TYPES SERVER" which has several possible valid values. This option setting specifies a list of encryption algorithms used for encrypting inbound and outbound network traffic for a DB instance. The OEM option has option settings for specifying a port and security groups for governing access to the OEM database.

Amazon RDS provides (i.e., via the Amazon RDS console, API, or CLI) the ability to create new option groups, copy existing option groups, modify existing option groups by adding or removing options, assign current values to option settings, and assign option groups to DB instances. Once an option group is assigned, the DB instance is empowered with the options as governed by the option settings.

Figure 2.2 extends Figure 2.1 by introducing additional entities and relationships focusing on option groups and options.

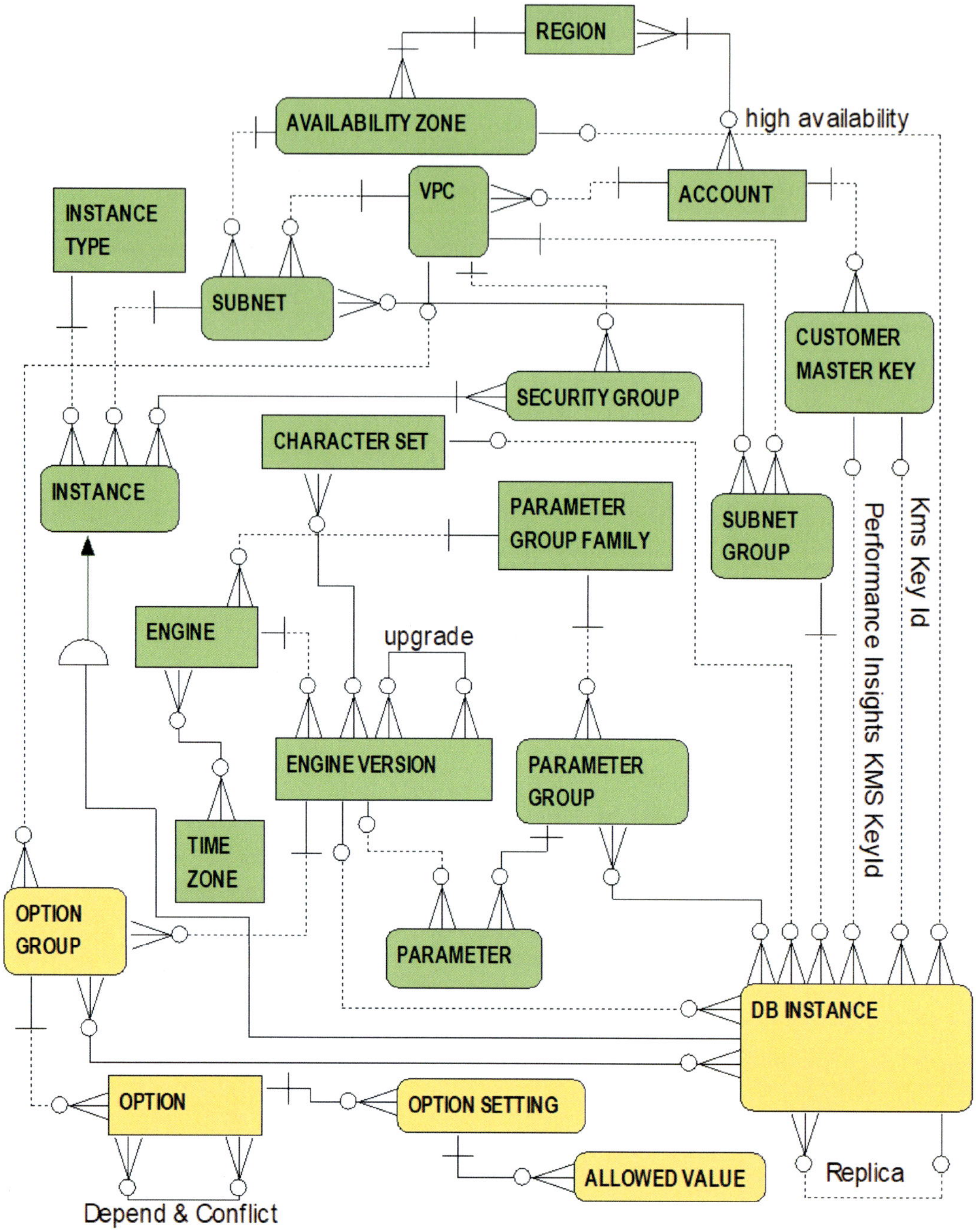

Figure 2.2 Option Group and Option Fundamentals

From Figure 2.2, the one-to-many relationship between the OPTION GROUP and OPTION entities indicates that each option group is composed of one or more options. The one-to-many relationship between the OPTION and OPTION SETTING entities indicates that each option has one or more option settings (e.g., as previously noted for Oracle, specifying the encryption algorithms for the “SQLNET.ENCRYPTION TYPES SERVER” option). The one-to-many relationship between the OPTION SETTING and ALLOWED VALUE entities indicate that each option setting has one or more valid values.

The one-to-many relationship between the ENGINE VERSION and OPTION GROUP entities indicates the major engine version applicable to an option group. An engine version can be the major engine version for many option groups.

The many-to-many reflexive relationship between the OPTION entity and itself suggests prerequisites and conflicts for each option. Each option can depend on, or be in conflict with, one or more other options. Each dependent (or conflicted) option can be a prerequisite for (or in conflict with) many other options. Recall from the explanation of a conceptual data model (CDM) and logical data model (LDM) in the Introduction, an LDM would resolve this many-to-many relationship (i.e., reveal more detail) by introducing a new *associative entity* (aka an *association class* in UML) including an attribute to flag (i.e., indicate) whether a related option is a prerequisite or conflict.[22]

The many-to-many relationship between the OPTION GROUP and DB INSTANCE entities shows that a DB INSTANCE can actually be associated with more than one option group. This acknowledges the possibility that RDS supports replacing the currently assigned option group with a second option group and that the status of this modification is pending. Most commonly, a DB instance has only one option group associated with it.

The one-to-many relationship between the VPC and OPTION GROUP entities indicates that each option group can be associated with a single VPC, meaning that the option group is applicable only to DB instances that reside within the associated VPC.

Event Notification

Life *events* happen for DB instances. Examples of events that can occur during the life cycle of a DB instance include:

- A DB instance is being shut down.
- A DB instance is being restarted after a previous shutdown.
- An error has occurred while restarting a DB instance.
- A backup has started (or has been completed).
- A parameter group for a DB instance is changing.
- Automatic backups for a DB instance are enabled (or disabled).
- DB instance storage has been consumed.

Amazon RDS enables the ability to configure *notifications* (e.g., email, text messages, calls to other AWS services, etc.) triggered when database life events transpire. For publishing such event notifications, Amazon RDS partners with yet another Amazon service, the Amazon Simple Notification Service (SNS).[23] Database life events are classified into *event categories* such as "backup", "failure", "low storage", "configuration", etc. It is also possible to set up a *subscription* for specific event categories, thereby sending notifications to subscribers whenever an event in any of the subscribed categories occurs. Consider, for example, the "backup" event category. If a subscription is setup to the "backup" category for a specific database instance, then notices will be generated for any backup start events and backup completion events that occur. The "low storage" category results in alerts for both storage being totally used up and also for 90% utilization.

Figure 2.3 extends Figure 2.1 by introducing additional entities and relationships focusing on event notifications.

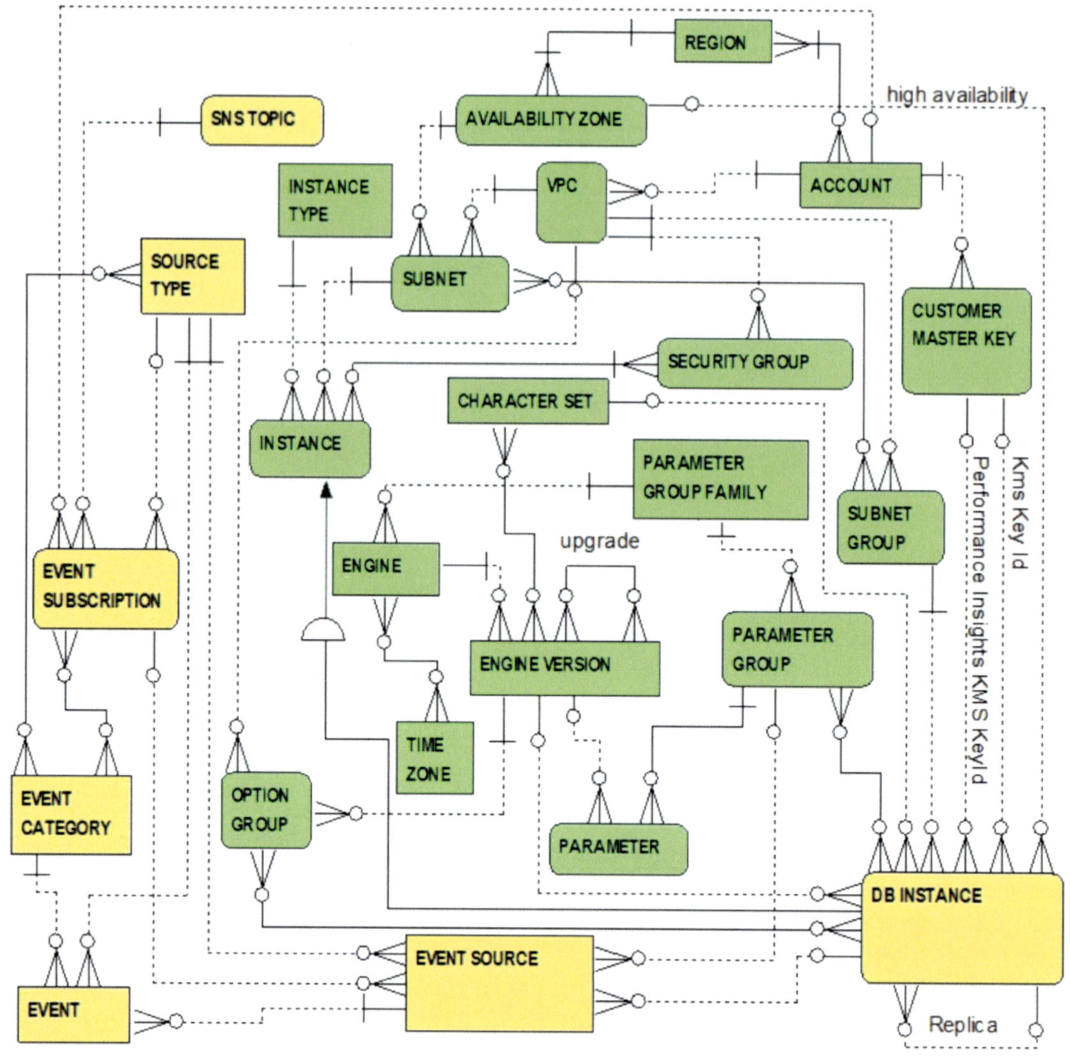

Figure 2.3 Event Notification Fundamentals

From Figure 2.3, the one-to-many relationship between the EVENT CATEGORY and EVENT entities indicates that each event category (e.g., "backup", "failure", "low storage", "configuration change", etc.) is associated with one or more events. The many-to-many relationship between the SOURCE TYPE and EVENT CATEGORY entities conveys that each event category is associated with many *source types* from which events can potentially be generated. In addition to DB instances, other types of event sources include parameter groups and database backups. Database backups are also referred to as *snapshots. Snapshots* will be further explained in a subsequent section. An example of events applicable to parameter groups includes, in the "configuration change" category, the occurrence of a parameter group modification (i.e., replacing the currently assigned parameter group by assigning a different parameter group to a DB instance). In the opposite direction, a source type can be associated with one or more event categories.

The one-to-many relationship between the SOURCE TYPE and EVENT SOURCE entities conveys that each source type is composed of many event sources such as specific DB instances and specific parameter groups. The one-to-many relationships from the DB INSTANCE and PARAMETER GROUP entities to the EVENT SOURCE entity shows that an event source is optionally either a specific DB instance or a specific parameter group.

The one-to-many relationship between the EVENT SOURCE and EVENT entities indicates that each event is generated from a specific source, in this case, either a DB instance or a parameter group.

The many-to-many relationship between the EVENT CATEGORY and EVENT SUBSCRIPTION entities conveys that each event subscription enables event notifications from possibly many event categories. But from which sources are these notifications? The one-to-many relationship between the EVENT SUBSCRIPTION and EVENT SOURCE entities indicates the specific sources (e.g., conceivably many DB instances) to which the subscription applies. The one-to-many relationship between the SOURCE TYPE and EVENT SUBSCRIPTION entities reveals that each subscription can be associated with a specific source type (e.g., if no event categories are associated). If a source type is specified (e.g., DB instances) but no event sources are specified (e.g., no specific DB instances) for an event subscription, then event notifications are received for all event sources belonging to the customer account for the specified source type (e.g., all DB instances). Specifying neither a source type nor an event source generates event notifications from all RDS sources (i.e., all DB instances, parameter groups, etc.) belonging to the customer account.

Note also the one-to-many relationship between the SNS TOPIC and EVENT SUBSCRIPTION entities revealing that each subscription specifies an SNS topic. Simply stated, an SNS topic is a named communication pathway enabling RDS to publish messages to SNS so that subscribers can receive event notifications. SNS topics and their associated *endpoints* (e.g., email addresses, text messages, HTTP/HTTPS entry points for web services, etc.) for receiving notifications are created via Amazon SNS (i.e., the Amazon SNS console, API, or software development kits) or via the Amazon RDS console.

Amazon SNS details (i.e., conceptual data modeling details) are beyond the scope of this immediate RDS conceptual modeling effort. It is envisioned that someday this paragraph -- in an updated edition -- will be modified to include a reference to a separate conceptual data model for Amazon SNS. Indeed, the world will be a much better place when all AWS services are graced with the accompaniment of their respective conceptual data models.

Reserved Database Instances

Amazon RDS is equipped with features offering AWS customers opportunities for substantial cost reductions associated with running database instances. One such offering is referred to as a *reserved instance.* A reserved instance is not the same as an actual running database instance. In essence, a reserved instance provides the prospect of (i.e., a reservation for) discounted pricing on one or more actually running or new database instances in exchange for a long term purchase commitment (i.e., reserving a DB instance for a one-year or three-year term).

So basically, an account purchases *reserved instance offerings* for price breaks. Each such reservation requests one or more DB instances having a specific instance type, a time period (i.e., the time duration of the reservation, either one year or three years), a payment option (e.g., a full upfront payment), the number of instances to be reserved, etc. The discount on usage charges varies depending on factors such as the instance type, region, database engine, and the selected payment option.

Reserved instance payment options include:

- *No Upfront:* For this payment option, no upfront advanced payment is involved. Instead, a discounted hourly rate is applied for the duration of the time period for each of the requested DB instances. This option is not allowed for a three year commitment (i.e., this option is only offered for a one-year time period).

- *Partial Upfront*: This option involves a less than full payment at the beginning of the reservation time period. A discounted hourly rate then applies for the term of the reservation for each of the requested instances, whether or not the reserved instances are actually used.

- *All Upfront:* For this option, the payment is completely made at the beginning of the reservation. No other charges apply for the duration of the reservation.

In general, launching a new DB instance in harmony with the specifications of a purchased reserved instance offering (e.g., same DB instance class and engine) enables Amazon RDS to apply the reduced price offering to the new DB instance. That is, for the duration of the reservation, if a DB instance is launched having the same configuration parameters as specified for the reservation, then the reduced pricing applies.

In general, newly launched DB instances whose configuration is not in harmony with a purchased reserved instance offering are billed at the prevailing on-demand rate. In certain cases, however, and for some database engines (e.g., MariaDB, MySQL, etc.), RDS does provide flexibility (i.e., referred to as *Size-Flexible Reserved Instances*) for applying reduced pricing to DB instances that are sized differently (i.e., but with the same region, engine, and instance family) from the offered reserved instance size.

Reserved instance discounts can also apply to currently running DB instances that are harmoniously or size-flexibly configured. Purchased reserved instances are automatically applied to the billing of active instances matching the reserved instance configuration. Charges for purchased reserved instances apply whether or not DB instances are actually being used.

Figure 2.4 extends Figure 2.3 by introducing additional entities and relationships focusing on reserved database instances.

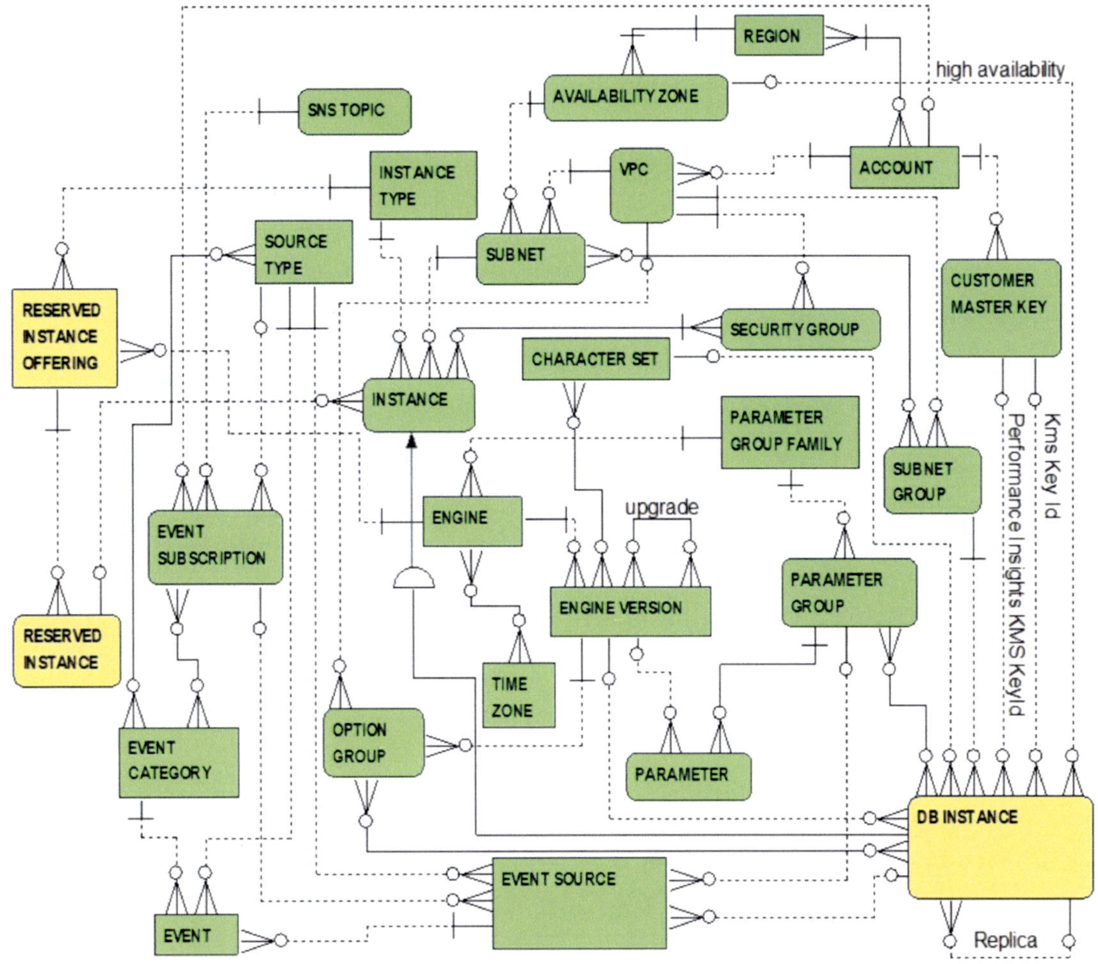

Figure 2.4 Reserved Database Instance Fundamentals

From Figure 2.4, the RESERVED INSTANCE OFFERING entity represents the starting point for purchasing reserved instances. The process starts with Amazon RDS providing a list of reserved instance offerings, when a user searches for offerings prior to actually purchasing reserved instances (e.g., via the Amazon RDS console). The one-to-many relationship from the INSTANCE TYPE entity to the RESERVED INSTANCE OFFERING entity indicates that an instance type (i.e., a DB instance class) can be referenced on one or more offerings and that each reserved instance offering identifies a single instance type. The one-to-many relationship from the ENGINE entity to the RESERVED INSTANCE OFFERING entity identifies the database engine for which the reserved instance offering applies. In addition to instance type and engine (i.e., foreign keys), noteworthy attributes of the RESERVED INSTANCE OFFERING entity include:

- *duration*: The time period length of the reserved instance commitment.
- *fixed price*: The reserved instance fixed purchase amount.
- *usage price*: The charge amount for each hour of reserved instance use.
- *offering type*: The payment option (i.e., “no upfront,” “partial upfront,” or “all upfront.”).
- *multi AZ*: Signifies whether or not the reserved instance offering pertains to multi-AZ deployments.
- *recurring charge amount*: The recurring monthly price for “no upfront” and “partial upfront” offerings.

Each reserved instance offering can spawn one or more reserved instances, as represented by the one-to-many relationship between the RESERVED INSTANCE OFFERING entity and the RESERVED INSTANCE entity. Beyond attributes inferred from its relationship to the RESERVED INSTANCE OFFERING entity, examples of significant attributes of the RESERVED INSTANCE entity include:

- *start time*: The reserved instance's date and time of commencement.
- *instance count*: The acquired number of DB instance reservations.
- *state*: Indicates the procurement status of the reserved instance. Status examples include "active", "retired", "payment-failed", "payment-pending", etc.

The one-to-many relationship between the RESERVED INSTANCE and INSTANCE entities suggests that each reserved instance (i.e., the discounted pricing associated with each reserved instance) can apply to one or more actually running database instances.

Database Backups

Amazon RDS provides automated backups of DB instances and also enables creation of manual (i.e., user initiated) backups. RDS backups, whether automated or user initiated, are referred to as *database snapshots*. RDS backs up the entire DB instance including all individual databases managed within the database instance. RDS enables both full backups and *incremental* backups whereby only the data modified since the previous snapshot is copied.

The RDS process for maintaining automated backups is governed by two previously noted DB INSTANCE attributes:

- *preferred backup window*: If automatic backups are enabled, this indicates the time range, a user-defined daily time window, when automated backups are created.
- *backup retention period*: The number of days that automatic backups are saved prior to being purged.

Figure 2.5 extends Figure 2.3 by introducing additional entities and relationships focusing on Amazon RDS backups.

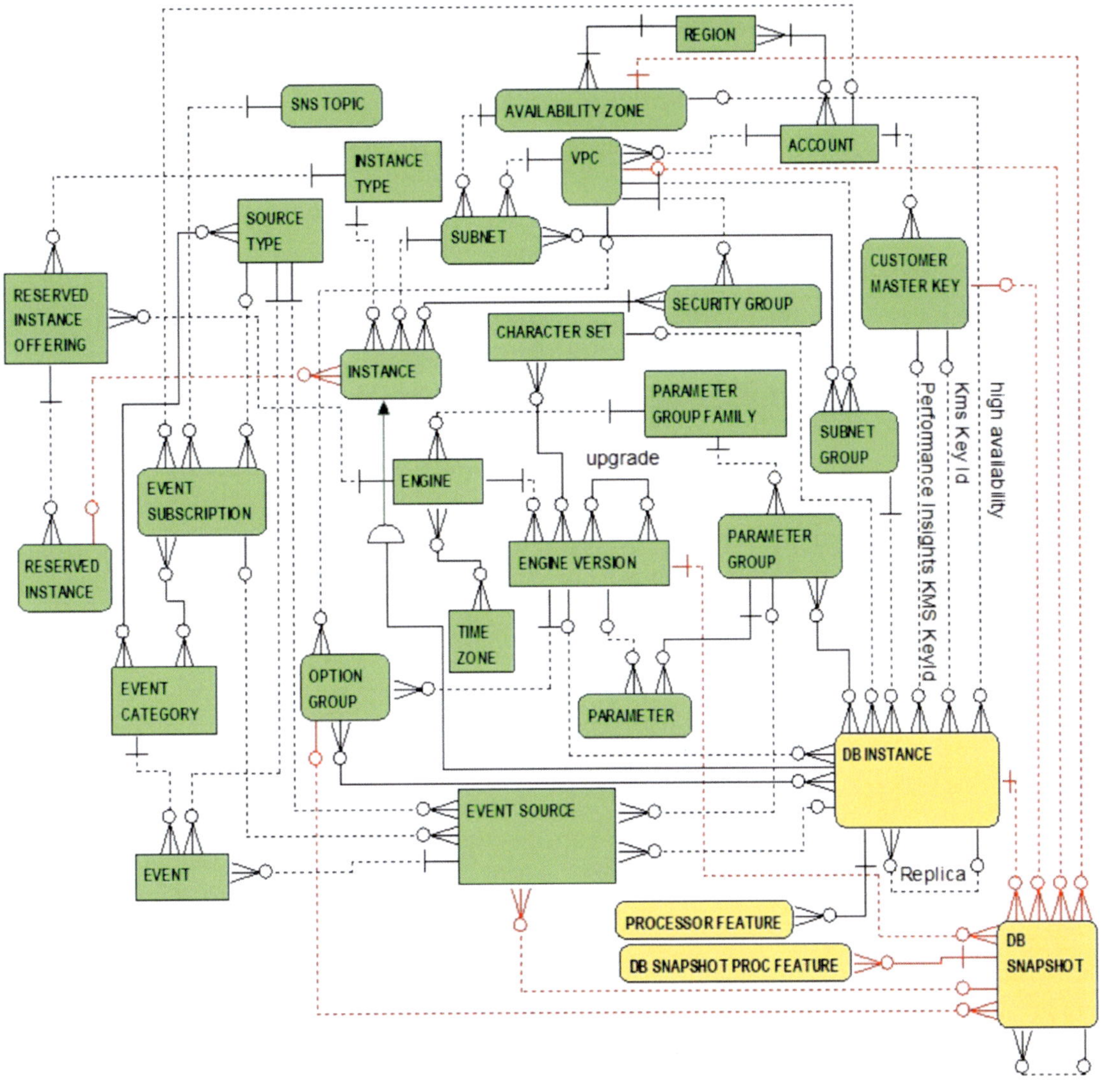

Figure 2.5 Database Backups

From Figure 2.5, the DB SNAPSHOT entity maintains backup information for both automated and manual backups. The one-to-many relationship between the DB INSTANCE and DB SNAPSHOT entities reveals that many DB snapshots can be maintained for each DB instance. By virtue of its potential need for database recovery, the DB SNAPSHOT entity duplicates crucial DB INSTANCE relationships necessary for accurately restoring a DB instance. These include one-to-many relationships from OPTION GROUP, CUSTOMER MASTER KEY, VPC, AVAILABILITY ZONE, and ENGINE VERSION entities to the DB SNAPSHOT entity. These relationships reflect the essential state of the DB instance at the time of the backup (e.g., the option group, encryption master key, VPC, and availability zone in effect at the time of the snapshot creation).

The one-to-many relationship from the DB SNAPSHOT entity to the EVENT SOURCE entity reflects that snapshots can also be the source for event notifications (i.e., in addition to parameter groups and DB instances). Note also the one-to-many relationship from the DB SNAPSHOT entity to the DB SNAPSHOT PROC FEATURE entity reflecting that snapshots also duplicate DB instance *processor features* in effect at the time of the backup as provided by the PROCESSOR FEATURE entity and its association with the DB INSTANCE entity. Valid processor features include the *core count* (i.e., number of CPU cores) and *threads per core* in effect for the DB instance at the time of the backup.

The one-to-many reflexive relationship from the DB SNAPSHOT entity to itself indicates that a DB snapshot can be copied from another DB snapshot (i.e., a DB snapshot can serve as the source for creating possibly many copies of the source DB snapshot).

In addition to identifiers for option group, customer master key, VPC, availability zone, engine version, and the source DB snapshot ID if copied (i.e., foreign keys), noteworthy attributes of the DB SNAPSHOT entity include:

- *allocated storage:* The storage size allocated for the snapshot expressed in gibibytes (GiB).
- *encrypted:* A boolean indicator flagging whether or not snapshot data is encrypted.
- *status:* Indicates the DB snapshot's completion status.
- *percent progress:* Indicates the progress of snapshot completion as a percentage of the total amount of DB instance data that has been moved to the snapshot.

Amazon RDS Database Logs and Log Types

Many database engines provide the ability to monitor ongoing database activity and collect monitoring data which is then preserved in database *log files*. Amazon RDS enables the monitoring of DB instances and collection of data in log files to assist in troubleshooting performance issues. Retrieval and analysis of performance data in log files also promotes an understanding of load conditions and traffic patterns, comparing historical data for contrast with current performance data, and detecting performance trends.

Examples of log files for the MariaDB engine are the *error log*, *slow query log*, *general log*, and *audit log*. Oracle log files include *alert logs*, *audit files*, *trace files*, and *listener logs*. Amazon RDS enables access to log files (i.e., for viewing, downloading, etc.) via the Amazon RDS console and API. Amazon RDS also collaborates with *Amazon CloudWatch Logs*, yet another AWS service, which provides access to log files generated by most database engines.[24] RDS database engines can be configured to export their log files to CloudWatch Logs. CloudWatch Logs then enables published log data to be analysed in real-time and provides the ability to create alarms and view metrics. Log files published to CloudWatch Logs can be saved for an unlimited length of time unless a specified retention period has expired. The Amazon RDS console enables choosing the logs to be published to Amazon CloudWatch Logs.

Figure 2.6 extends Figure 2.3 by introducing additional entities and relationships focusing on Amazon RDS database logs and log types.

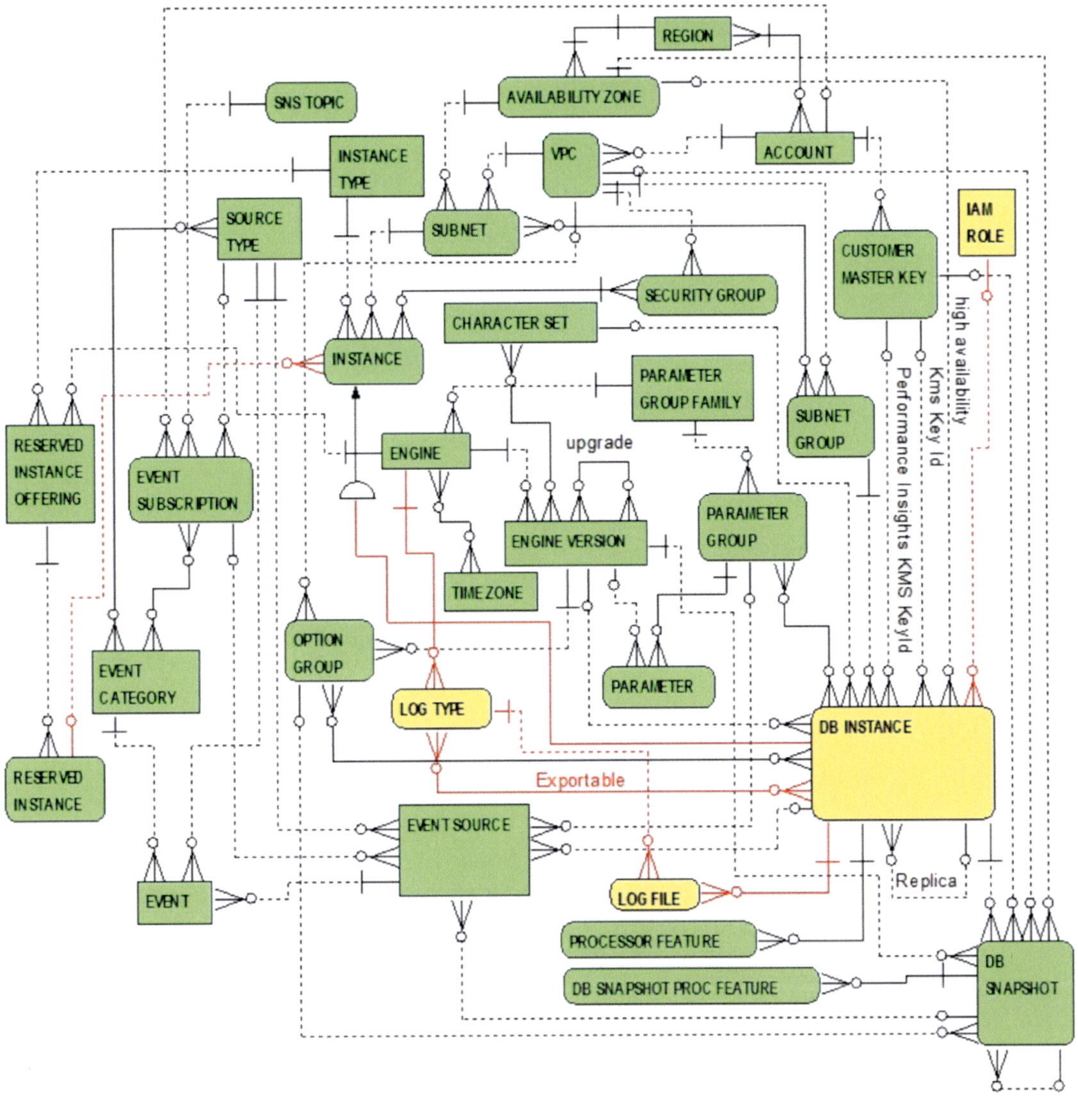

Figure 2.6 RDS Database Logs and Log Types

From Figure 2.6, the one-to-many relationship between the ENGINE and LOG TYPE entities reveals that each database engine can support multiple log types. As previously noted, the Oracle engine for example supports *alert*, *audit*, *trace*, and *listener* log types. The many-to-many relationship between the LOG TYPE and DB INSTANCE entities acknowledges that a DB instance can export many of its engine's log types to Amazon CloudWatch Logs. Each log type of course can be exported by many DB instances. The one-to-many relationship between the DB INSTANCE and LOG FILE entities reveals that each database instance can generate multiple log files. The one-to-many relationship between the LOG TYPE and LOG FILE entities indicates that each log file is associated with a valid log type. All three RDS user interfaces (i.e., the Amazon RDS console, AWS CLI, and Amazon RDS API) enable access to database log files (i.e., for viewing, downloading, etc.).

Note also the one-to-many relationship between the IAM ROLE and DB INSTANCE entities. IAM refers to AWS Identity and Access Management, another AWS service, which enables customers to maintain authorization information necessary for accessing other AWS services.[25] In this case, the IAM role associated with the DB instance permits

RDS to send *enhanced monitoring metrics* (e.g., CPU and memory utilization for RDS processes and OS processes) to Amazon CloudWatch Logs.

Chapter Summary

Figure 2.7 summarizes RDS essential entities and relationships.

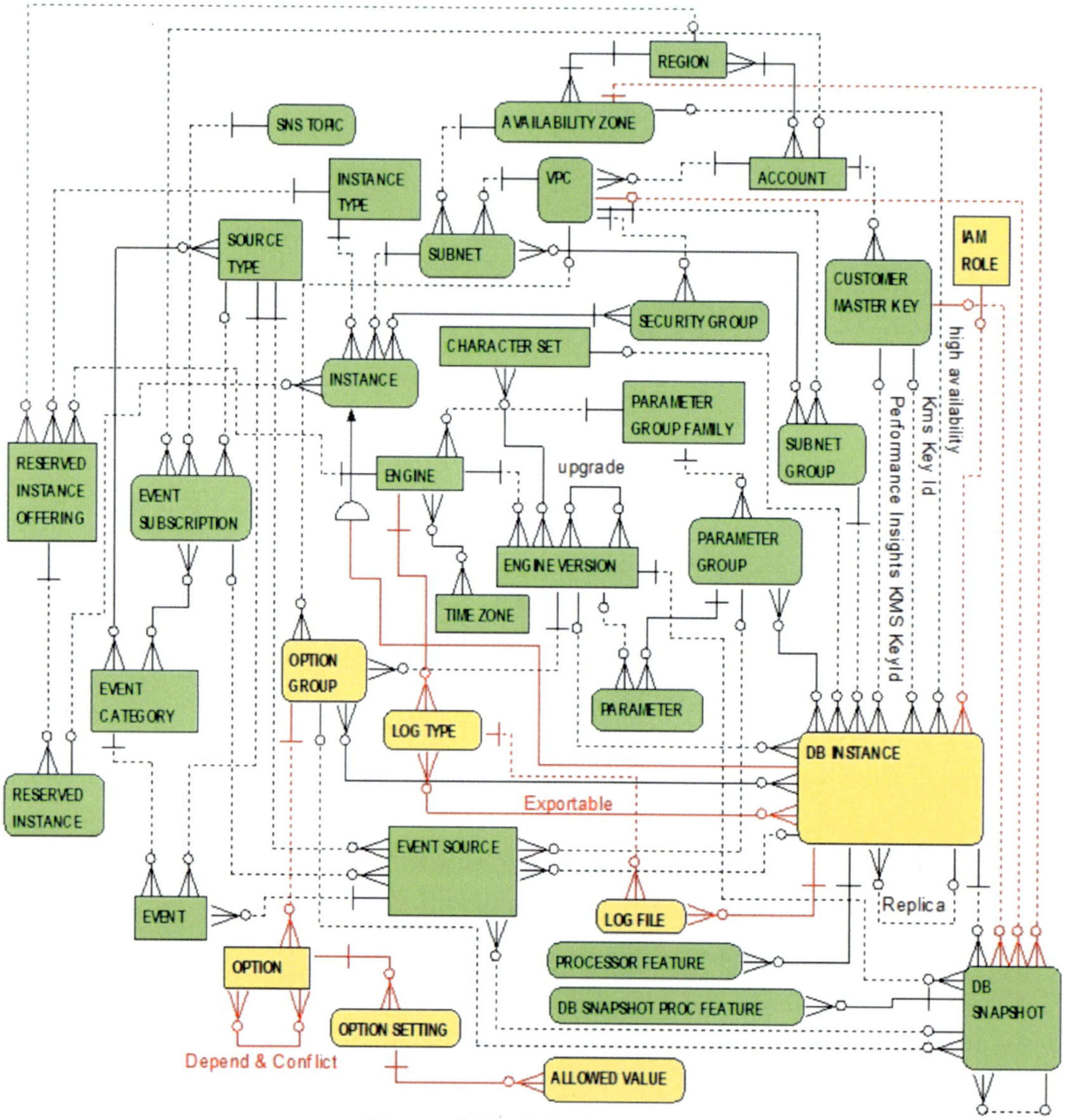

Figure 2.7 RDS Summary

RDS Fundamentals: A variety of relational DBMS engines and versions of database engines are supported by RDS (Oracle DB, MySQL, PostgreSQL, Microsoft SQL Server, and MariaDB). Each version of a database engine can support many character sets, one of which can be designated as the default character set. A character set (e.g., UTF8) can be supported by many engine versions. Engine versions that are possible valid targets for upgrade. More specifically, each engine version can possibly be upgraded to many later and greater versions, and an engine version can be the target of many prior engine versions.

A parameter group is a collection of many parameters used for setting up a database. Values are assigned to database parameters such as for time zone, character set, optimization features, audit trails, table cache size, maximum number of processes for parallel execution, categories of database memory (e.g., Oracle's system global area, SGA). Each parameter can be associated with an initial engine version (i.e., the initial engine version for which the parameter is relevant).

A database engine can support many time zones and a time zone can be supported by many engines. For example, UTC is the default time zone for RDS MySQL and MariaDB databases. For these engines, the *time_zone* parameter in a parameter group can be set to the specific regional time zone corresponding to the application's time zone.

An RDS DB instance is *a kind of* EC2 instance for running isolated database environments. As *a kind of* EC2 instance, each DB instance inherits an instance type, resides in a subnet, is associated with security groups, etc. Depending on the database engine, each DB instance can support a single database (e.g., an Oracle database with multiple *schemas*) or multiple databases (e.g., a Microsoft SQL Server environment). Each DB instance is associated with a single DBMS engine version and a single character set.

Each DB instance can be associated with an alternate availability zone for implementing high availability. RDS provides an option for having a secondary DB instance synchronized with the primary DB instance for failover to the secondary DB instance in this alternate availability zone.

Amazon encryption is provided by the AWS Key Management Service (AWS KMS), which uses *customer master keys* (CMKs) as the basis for encrypting or decrypting data. Two CMKs are optionally provided (as foreign keys, FKs) to a DB instance -- one is used for encrypting DB instance data, and a second CMK encrypts *Performance Insights data* for troubleshooting performance problems.

A *subnet group* within a VPC is a collection of subnets and each subnet can belong to many subnet groups. Each DB instance is associated with a single subnet group for assigning the DB instance to an appropriate subnet. Each subnet group should have subnets dispersed among multiple availability zones to facilitate high availability.

A DB instance can be associated with more than one parameter group (i.e., the currently assigned parameter group can be replaced with a second *pending* parameter group). A DB instance normally has only one parameter group associated with it.

A primary DB instance serves as the source of data copied to one or more DB instances serving as *read replicas* (i.e., applicable to MariaDB, MySQL and PostgreSQL). A read replica provides read-only access and alleviates the load on the primary DB instance. Any data updates on the primary DB instance are automatically and asynchronously pushed to its read replicas.

Option Groups and Options: An option group is composed of one or more options with each option having one or more option settings (e.g., Oracle's "SQLNET.ENCRYPTION TYPES SERVER" option). Each option setting has one or more valid values that are allowed. An engine version can be the major engine version for many option groups.

Each option can depend on, or be in conflict with, one or more other options. Each dependent (or conflicted) option can be a prerequisite for (or in conflict with) many other options. A DB instance can be associated with more than one option group (i.e., the currently assigned option group is replaced with a second *pending* option group). An option group is associated with a single VPC and is eligible only to DB instances in the associated VPC.

Event Notification: For DB instances, life *events* happen that fall into *event categories* (e.g., "backup", "failure", "low storage", "configuration change", etc.). Each event category is associated with many *source types* from which event notifications can potentially be generated. In addition to DB instances, other types of event sources include parameter groups and database backups (aka *snapshots)*. A source type can be associated with many event categories. Each source type is composed of many *event sources* such as specific DB instances, parameter groups, or snapshots. An event source is optionally either a specific DB instance, parameter group, or snapshot. Each event is generated from a source.

An *event subscription* enables event notifications from possibly many event categories. An *event subscription* also indicates the specific sources (e.g., many DB instances) to which the subscription applies. Each event subscription can optionally be associated with a specific source type. If a source type is specified for an event subscription (e.g., the subscription is for DB instances) but no event sources are specified (i.e., no specific DB instances are specified) then event notifications are received for all RDS sources of the specified source type (i.e., all DB instances) belonging to the customer account. Specifying neither a source type nor an event source generates event notifications from all RDS sources (i.e., all DB instances, parameter groups, etc.) belonging to the customer account.

Reserved Database Instances: Amazon RDS enables the purchase of *reserved instances* -- by providing a list of *reserved instance offerings*. An instance type (i.e., a DB instance class) can be referenced on one or more offerings and each reserved instance offering identifies a single instance type. Each reserved instance offering applies to a single database engine. Each reserved instance offering purchased by an account can spawn one or more reserved instances. Each reserved instance (i.e., the discounted pricing associated with each reserved instance) can apply to one or more actually running DB instances.

Database Backups: *DB snapshots* maintain backup information for both automated and manual backups. Many DB snapshots can be maintained for each DB instance. Each DB snapshot duplicates crucial DB instance relationships necessary for accurately restoring a DB instance. These relationships reflect the essential state of the DB instance at the time of the backup (e.g., the option group, encryption master key, VPC, availability zone, etc.) in effect at the time of the snapshot creation). In addition to parameter groups and DB instances, snapshots can also be the source of event notifications. Snapshots also duplicate DB instance *processor features* in effect at the time of the backup. Valid processor features include the *core count* (i.e., number of CPU cores) and *threads per core* in effect for the DB instance at the time of the backup. A DB snapshot can serve as the source for creating possibly many snapshot copies.

Logs and Log Types: Each database engine can support multiple log types (e.g., Oracle supports *alert*, *audit*, *trace*, and *listener* log types). A DB instance can export many of its engine's log types to Amazon CloudWatch Logs. Each log type can be exported by many DB instances. Each DB instance can generate multiple log files. Each log file is associated with a valid log type. The Amazon RDS console, AWS CLI, and Amazon RDS API enable access to database log files (i.e., for viewing, downloading, etc.). IAM (AWS Identity and Access Management) enables customers to maintain user authorization information necessary for accessing other AWS services. The IAM role associated with the DB instance permits RDS to send *enhanced monitoring metrics* (e.g., CPU and memory utilization for RDS processes and OS processes) to Amazon CloudWatch Logs.

Chapter 3 – Amazon Aurora

Amazon Aurora Introduction

Amazon Aurora (Aurora) represents another option, in addition to Amazon RDS, for deploying relational database instances in AWS. Aurora, accessed via Amazon RDS, is a managed relational database engine and is compatible with both MySQL and PostgreSQL. Existing MySQL or PostgreSQL applications can migrate their databases to Aurora without the need for application conversion or modification.

Other motives for moving to Aurora, in addition to Aurora's support for familiar open source RDBMS engines, are its prospect for substantially improved performance and high availability. Specifically, the Aurora database engine is equipped with *solid-state storage devices (SSD, i.e., solid-state drives; aka solid-state disks)* deliberately engineered to significantly improve database response time and throughput. SSDs, unlike hard disk drives (HDDs), have no spinning disks or other moving electromechanical components, and offer a higher degree of fault-tolerance. As a managed service, Aurora oversees the need for automatically expanding storage requirements -- up to a maximum of 64 terabytes -- without service disruption (i.e., without the need for scheduling planned downtime).

High availability is also achieved by *database clustering* -- Aurora's ability to configure and administer a *cluster* of DB instances. An Aurora cluster consists of a *primary* DB instance -- responsible for read and write operations -- and up to as many as 15 *read replica* DB instances, whose updates are automatically synchronized with the primary DB instance. Aurora read replica DB instances (aka replicas) are deployed for offloading data retrieval requests and also serve as possible targets for failover in the event of primary DB instance failure. Aurora can automatically switch to one of the read replica database instances in the event that the primary DB instance is deemed irreparable. Of course, high availability is further promoted by scattering the primary instance and its replicas over multiple availability zones.

Performance, availability and reliability are also enhanced because cluster data is maintained within a single, virtual *cluster volume* replicated across three availability zones, mutually accessible by all DB instances within the cluster.

Figure 3.1 introduces entities and relationships focusing on Aurora's fundamental concepts.

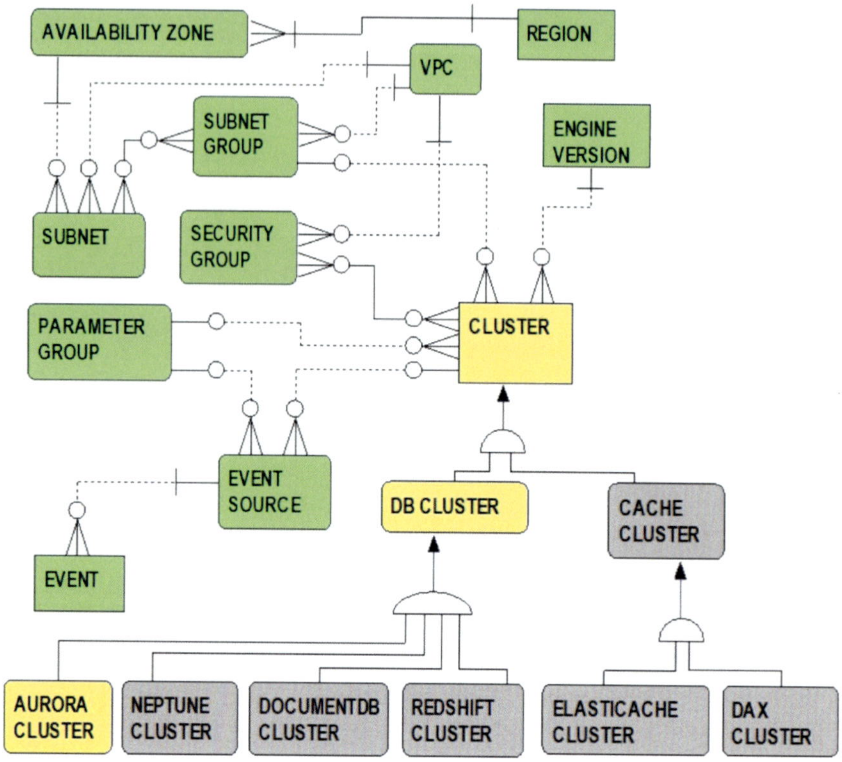

Figure 3.1 Entities and Relationships … Aurora Fundamental Concepts

From Figure 3.1, the CLUSTER entity introduces Aurora's ability to manage a cluster of DB instances. The CLUSTER entity is a high level *super-type* that represents all clusters (i.e., groups) of instances for all Amazon database services in general. In addition to Aurora clusters, there are other sub categories of clusters that will be explored in subsequent chapters devoted to other Amazon database services. Specifically, as illustrated in Figure 3.1, the CLUSTER super-type is made up of two *sub-types* represented by the DB CLUSTER and CACHE CLUSTER entities. The CACHE CLUSTER entity will be further explored in Chapter 7 (DynamoDB clusters) and Chapter 8 (ElastiCache clusters).

The DB CLUSTER and CACHE CLUSTER entities, as sub-types, inherit attributes and relationships from their super-type CLUSTER entity. The CACHE CLUSTER entity represents the set of all *cache clusters*. Each cache cluster is *a kind of* cluster. The DB CLUSTER entity represents the set of all *DB clusters*. Each DB cluster is *a kind of* cluster.

Figure 3.1 illustrates that the DB CLUSTER entity is also a super-type for four additional sub-types: the AURORA CLUSTER, NEPTUNE CLUSTER, DOCUMENTDB CLUSTER, and REDSHIFT CLUSTER entities. Each of these sub-types inherits attributes and relationships from the DB CLUSTER entity. The AURORA CLUSTER entity is of course the immediate focus of this chapter. The NEPTUNE CLUSTER entity, DOCUMENTDB CLUSTER entity, and REDSHIFT CLUSTER entity are respectively explored in Chapters 4, 5 and 6. Aurora, Neptune, DocumentDB and Redshift clusters are each *a kind of* DB cluster.

Relationships to the CLUSTER entity reveal that every cluster -- whether a DB cluster or cache cluster -- can be associated with many security groups, a single subnet group, a single parameter group, and a specific engine version. This means that a cluster of instances -- either DB instances or cache instances -- has:

- A set of security groups for controlling access to cluster instances (i.e., DB instances or cache instances).
- A subnet group within whose subnets cluster instances can be launched.
- A single parameter group specifying parameters for configuring all cluster instances.
- The engine version (DB engine version or cache engine version) for all cluster instances.

Note also the one-to-many relationship from the CLUSTER entity to the EVENT SOURCE entity emphasizing that any cluster -- whether a DB cluster or cache cluster -- can serve as a source for event notifications.

In addition to identifiers for engine version, subnet group, and parameter group (i.e., foreign keys), noteworthy sample attributes of the CLUSTER entity -- and, therefore, inherited by all CLUSTER sub-types -- include:

- *cluster identifier:* A user-supplied cluster identifier.
- *cluster create time:* The time when the cluster was created.
- *status:* Indicates the current state of the cluster.
- *endpoint:* Specifies the cluster's access information (i.e., DNS host name and port number) enabling applications to connect to the cluster (i.e., rather than connecting to a specific DB instance in the cluster).
- *preferred maintenance window:* Identifies the time period when maintenance activities (e.g., system upgrades) are preferred.

DB Clusters

Figure 3.2 extends Figure 3.1 by focusing on entities and relationships for clarifying and differentiating DB clusters.

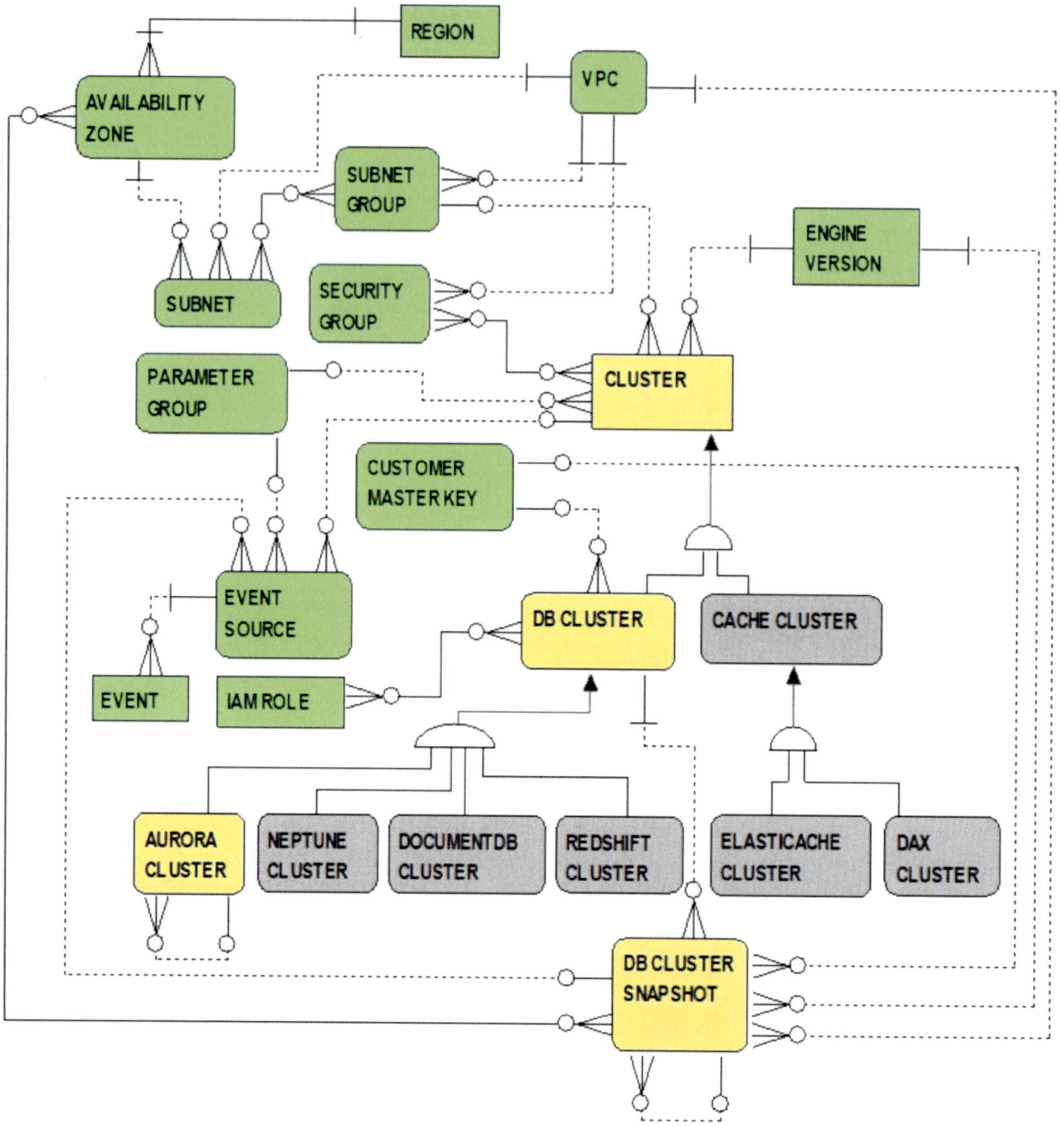

Figure 3.2 Entities and Relationships Peculiar to DB Clusters

From Figure 3.2, the DB CLUSTER entity represents the set of all Aurora, Neptune, DocumentDB, and Redshift clusters. In addition to inheriting higher level attributes and relationships from its super-type CLUSTER entity (i.e., many security groups, a single subnet group, a parameter group, an engine version, and as a source for event notifications), the DB CLUSTER entity provides additional relationships and attributes common to all DB clusters (Aurora, Neptune, DocumentDB, and Redshift clusters).

The one-to-many relationship from the CUSTOMER MASTER KEY entity to DB CLUSTER entity indicates that all DB clusters (i.e., data maintained by all instances in DB clusters) are optionally encrypted using a KMS customer master key. The many-to-many relationship between the DB CLUSTER entity and the IAM ROLE entity indicates that each database cluster can possibly have one or more IAM roles, thereby enabling access -- by Aurora, Neptune, Document DB, and Redshift -- to other AWS services on behalf of a DB cluster.

The one-to-many relationship from the DB CLUSTER entity to DB CLUSTER SNAPSHOT indicates that a DB cluster (i.e., all instances associated with a DB cluster) can be backed up by one or more *DB cluster snapshots*. Similar to the DB INSTANCE SNAPSHOT entity (described in Chapter 2), the DB CLUSTER SNAPSHOT entity duplicates crucial DB CLUSTER and CLUSTER relationships in order to accurately restore a DB cluster if necessary. These duplicated relationships include one-to-many relationships from CUSTOMER MASTER KEY, VPC and ENGINE VERSION entities to the DB SNAPSHOT entity. These relationships reflect the essential state and condition of the DB cluster at the time of the backup (i.e., the encryption key, the VPC, and engine version in effect at the time of DB cluster snapshot creation). The many-to-many relationship between the DB CLUSTER SNAPSHOT entity and the AVAILABILITY ZONE entity indicates the availability zones within which DB instances backed up in the DB cluster snapshot can be restored.

The one-to-many relationship from the DB CLUSTER SNAPSHOT entity to the EVENT SOURCE entity indicates that a DB cluster snapshot can serve as yet another source for event notifications (e.g., in addition to DB instances, parameter groups, DB snapshots and DB clusters). The one-to-many reflexive relationship from the DB CLUSTER SNAPSHOT entity to itself indicates that a DB cluster snapshot can be copied from another DB cluster snapshot (i.e., a DB cluster snapshot can serve as the source for creating possibly many copies of the source DB cluster).

In addition to attributes inherited from its super-type CLUSTER entity, and the customer master key (i.e., foreign key), noteworthy sample attributes of the DB CLUSTER entity include:

- *backup retention period:* The number of days that automatic backups are saved prior to being purged.
- *database name:* Identifies the name of the DB cluster's initial database.
- *master username:* The master username for logging into the cluster's DB instances.
- *storage encrypted:* Indicates whether or not the cluster's data is encrypted.

Aurora Clusters

Figure 3.3 extends Figure 3.2 by distinguishing special relationships for Aurora clusters, a subset of DB clusters.

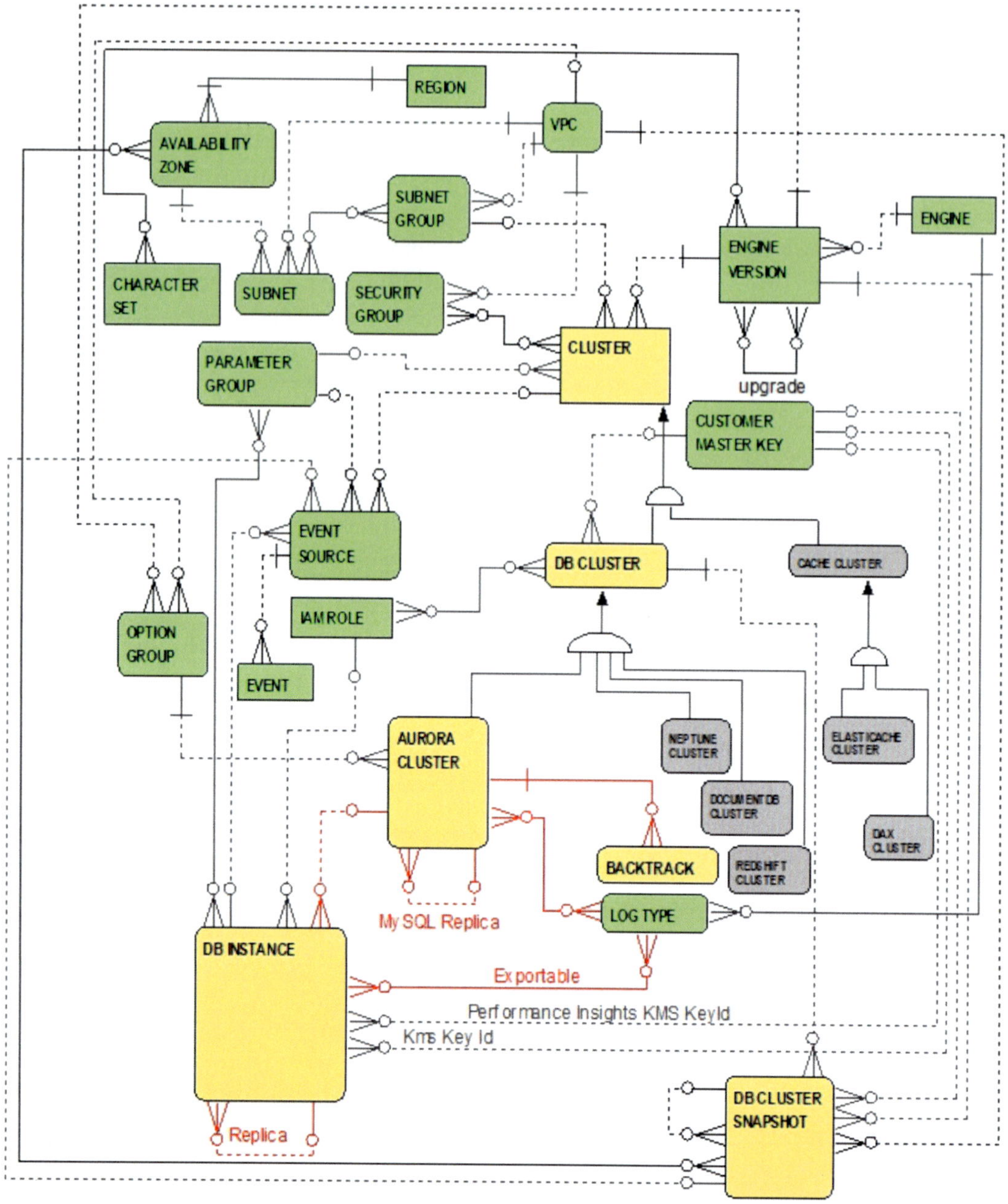

Figure 3.3 Entities and Relationships Peculiar to Aurora Clusters

From Figure 3.3, the AURORA CLUSTER entity, shares the similarities inherited from the DB CLUSTER and CLUSTER entities in its super-type hierarchy. Specifically, the AURORA CLUSTER entity inherits …

- Higher level relationships to its super-type DB CLUSTER entity (i.e., a KMS customer master key for data encryption, IAM roles, and DB cluster snapshots) and DB CLUSTER attributes.
- Higher level relationships to its grandparent CLUSTER entity (i.e., many security groups, a source for event notifications, a single subnet group, a parameter group, and an engine version) and CLUSTER attributes.

Figure 3.3 also celebrates additional relationships peculiar to the AURORA CLUSTER entity. The one-to-many relationship from AURORA CLUSTER entity to the DB

INSTANCE entity indicates that each AURORA cluster is composed of one or more DB instances (i.e., all DB instances running the same version of MySQL or all DB instances running the same version of PostgresSQL). One DB instance is primary and the remaining servers are read replicas. For example (via the Aurora console), when creating an Aurora cluster, Amazon RDS automatically creates the primary DB instance for the Aurora cluster (i.e., for reading, writing and modifying data to the cluster volume) and optionally creates a replica DB instance in a separate availability zone. More Aurora replicas can be added to an Aurora cluster using the web console, CLI, or API. Once an Aurora cluster is created, customary DBMS admin utilities (i.e., standard MySQL or PostgreSQL utilities) are used to connect to a database.

The one-to-many reflexive relationship from the DB INSTANCE entity to itself indicates the possibly that many DB instances can serve as read replicas for a primary DB instance (i.e., a primary DB instance is synchronized with possibly many read replicas).

The one-to-many reflexive relationship from AURORA CLUSTER entity to itself highlights that an Aurora MySQL cluster can be synchronized to one or more MySQL clusters serving as *Read Replica clusters* in separate regions from the source region. This improves read performance to more remote customers, and enormously facilitates application migration and disaster recovery. Such read replica clusters are referred to as *cross-region* DB clusters. This relationship is annotated to emphasize its applicability only to Aurora MySQL clusters.

The one-to-many relationship from AURORA CLUSTER entity to the BACKTRACK entity introduces the concept of *backtracking* supported for AURORA MySQL clusters. Without requiring a backup snapshot, backtracking enables an AURORA cluster to be rewound to an earlier time in history. Backtracking in essence performs an *undo* (e.g., for quickly recovering from or correcting blunders such as inadvertently dropping a table, or for undoing a duplicated data load -- aka *deduping*) with a minimum of service disruption.

If backtracking is enabled, Aurora archives *change records* (i.e., saved as modifications applied to the Aurora cluster) as the basis for performing backtracks. Aurora maintains a history of possibly many backtracks requested for an AURORA cluster including the date and time on which the cluster was backtracked, the current status of the backtrack, etc. Backtracking contrasts with performing a DB cluster snapshot, recovery from which requires creating a new cluster and usually involves a substantially longer recovery time.

The many-to-many relationship between the AURORA CLUSTER and LOG TYPE entities indicates the types of log files that can be published to Amazon CloudWatch Logs for all DB instances in the Aurora cluster. The LOG TYPE entity was previously addressed in Chapter 2. The one-to-many relationship from the OPTION GROUP entity to the AURORA CLUSTER entity reflects that a default option group is associated with each Aurora cluster and is thereby initially applicable to all DB instances in the Aurora cluster.

Note also that the one-to-many relationship from the AURORA CLUSTER entity to the DB INSTANCE entity is an optional relationship implying the possibility that an AURORA cluster can exist with zero DB instances provisioned for it. This feature, available for a subset of MySQL and PostgreSQL versions, is referred to as an Aurora *serverless* DB cluster as opposed to a *provisioned* cluster for which a primary DB instance and optional replica DB instances are specified. For Aurora *serverless DB* clusters, Aurora automatically manages compute and network resources (i.e., using proxy servers and *warm pools* of DB instances), and storage requirements. Aurora dynamically tweaks (up or down) compute and memory capacity to accommodate vicissitudes in application usage and actual workload (e.g., based on CPU utilization). Similar to provisioned Aurora clusters, the virtual storage volume for Aurora serverless DB clusters automatically scales to a maximum of 64 terabytes. Clearly, serverless clusters substantially simplify managing the capacity for applications with widely fluctuating workloads.

In addition to previously described attributes inherited from its super-type hierarchy (i.e., foreign keys and other attributes inherited from the CLUSTER and DB CLUSTER entities), noteworthy examples of attributes peculiar to the AURORA CLUSTER entity include:

- *backtrack consumed change records*: The count of backtrack change records retained for the cluster.
- *backtrack window*: Amount of time in seconds for which a backtrack may be wanted, (e.g., a target backtrack window of 48 hours).
- *earliest backtrack time*: The earliest time for which a DB cluster backtrack can be requested (e.g., based on accumulated change records and available storage).
- *deletion protection*: A flag indicating if the DB cluster is shielded from being deleted.
- *DB engine mode*: Indicates the engine mode (e.g., serverless, provisioned).
- *max capacity*: For a serverless cluster, the maximum capacity in terms of *Aurora Capacity Units* (ACUs) expressing the peak capacity (i.e., compute and memory) to which Aurora can automatically expand.

Additional attributes apply to the AURORA CLUSTER entity that are not peculiar and (as will be revealed in the next chapter) are also applicable to Neptune clusters, but not applicable to Redshift clusters. These additional attributes are:

- *multi AZ*: Indicates high availability in that DB instances are provisioned in more than one availability zone.
- *preferred backup window:* If automatic backups are enabled, this indicates the time range (i.e., a user-defined daily time window) when automated backups are created.
- *earliest restorable time:* For restoring a cluster to a specific point-in-time, the earliest possible time in the past to the cluster can be recovered.
- *latest restorable time:* For restoring a cluster to a specific point-in-time, the latest possible time in the past the cluster can be recovered.
- *allocated storage*: Always set to 1 for Aurora clusters indicating that Aurora storage size is not fixed but automatically scales as necessary.

Chapter Summary

Figure 3.4 extends Figure 3.3 and integrates additional relationships (from Chapter 2) for DB instances.

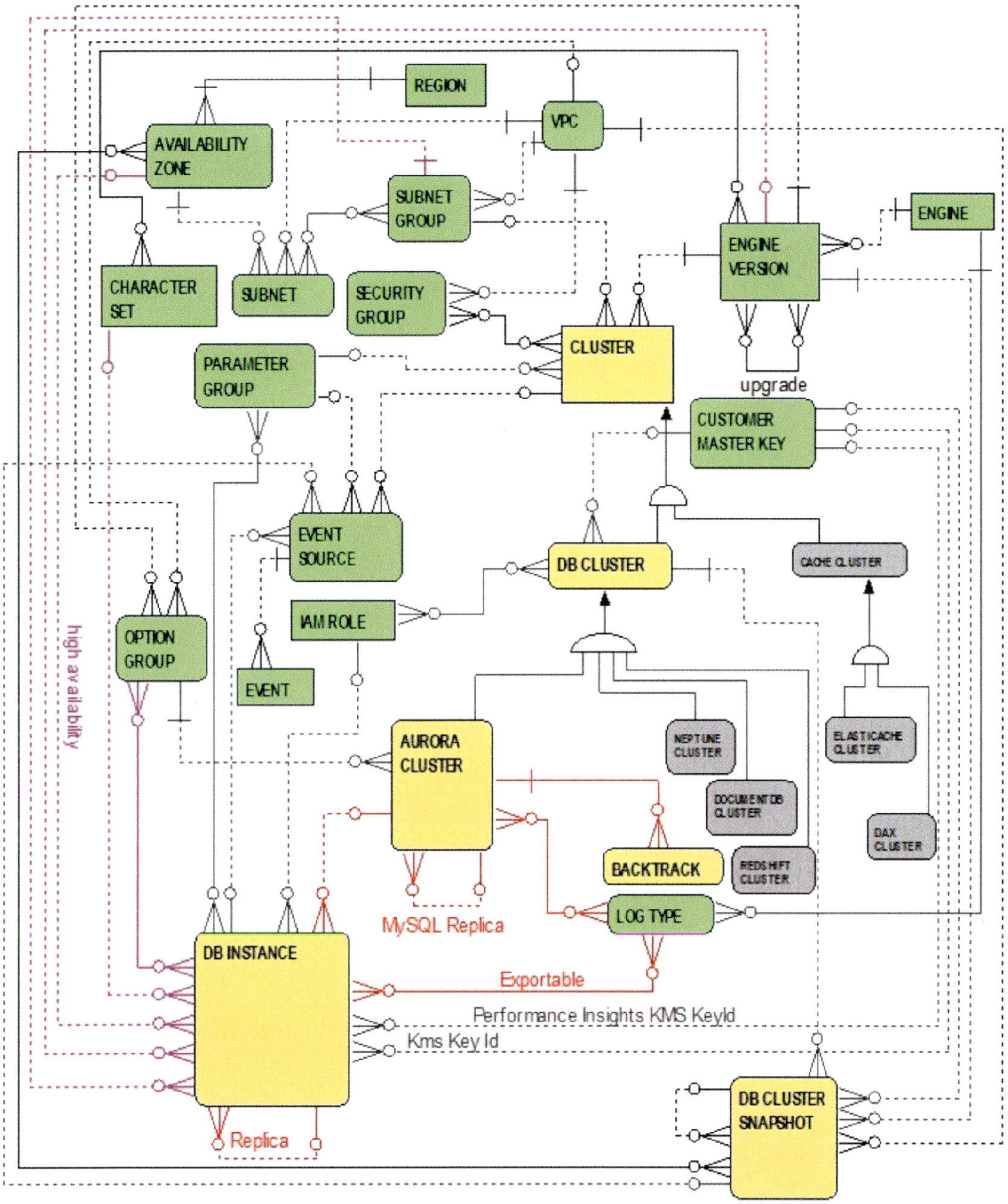

Figure 3.4 Aurora Summary

Aurora is a managed relational database engine and is compatible with both MySQL and PostgreSQL -- open source relational engines. Aurora features solid-state storage devices (SSD) for improved response time and throughput, and automatically expands storage capacity without service disruption.

Aurora introduces the concept of *database clustering* -- Aurora's ability to configure and administer a cluster of DB instances. For AWS database services, there is an *inheritance hierarchy of clusters* starting with two sub-types of clusters: *DB clusters* and *Cache clusters*. An Aurora cluster is *a kind of* DB cluster which in turn is *a kind of* cluster. An Aurora cluster inherits a KMS customer master key (i.e., a CMK for data encryption), IAM roles, and cluster snapshots because it is *a kind of* DB cluster. An Aurora cluster indirectly inherits (i.e., because a DB cluster is *a kind of* cluster) many security groups, a subnet group, a parameter group, an engine version, and is a source for event notifications. An Aurora cluster consists of a *primary* DB instance (for read and write operations) and as many as 15 *read replica* DB instances (updates are automatically synchronized from the primary). Aurora replicas offload data retrieval requests. Aurora automatically switches to a read replica if the primary DB instance crashes. Aurora maintains a single, virtual cluster storage volume (replicated across availability zones) which is accessible to all cluster instances.

An Aurora cluster for MySQL can be synchronized to one or more clusters serving as *read replica DB clusters* (aka *cross-region* DB clusters) in separate regions from the source region (improves read performance for remote customers, and expedites app migration and disaster recovery).

Aurora supports *backtracking* for MySQL clusters. Backtracking enables cluster data to be rewound to an earlier time in history (i.e., an *undo* for recovering from inadvertently dropping a table) with a minimum of service disruption. If backtracking is enabled, Aurora archives *change records* as the basis for performing backtracks. Aurora maintains a history of possibly many backtracks requested for the cluster including the date and time at which the DB cluster was backtracked, the current status of the backtrack, etc. Backtracking differs from restoring a cluster from a snapshot, which requires creating a new cluster and lengthier recovery time.

An AURORA cluster can support one or more log types indicating the types of log files that can be published (by all DB instances in the cluster) to Amazon CloudWatch Logs. A default option group is associated with each Aurora cluster -- initially applicable to all DB instances in the cluster.

Chapter 4 – Amazon Neptune

Amazon Neptune Introduction

Amazon Neptune (Neptune) is a *graph database* engine and -- like RDS and Aurora -- a managed AWS database service.[26] Routine database admin tasks, typically performed by DBAs, such as provisioning hardware (i.e., scaling additional compute, memory, and storage resources), performing backups, applying software patches, detecting and reacting to failures, and recovering databases from backups are expedited and often automated by Neptune. Like Amazon Aurora, Neptune is also equipped with solid-state virtual storage facilities which are optimized for database workloads and deliberately engineered to significantly improve database performance, reliability, and availability. Neptune also promotes high availability by enabling automatic failover to read replicas and data replication across availability zones.

A Neptune graph database is fundamentally different from SQL-based relational database approaches, such as those engines supported by RDS. As previously noted, the data manipulation language (DML) for accessing most relational databases is some version or adaptation of SQL (e.g., an ANSI standard SQL version or a vendor specific SQL version).

In contrast, Neptune is categorized as a NoSQL (i.e., "Non SQL" or "Not Only SQL") database service which essentially means that Neptune provides a non-relational approach to managing, querying and manipulating data.[27] Instead of abiding by familiar relational concepts (i.e., tables with columns and foreign keys for implementing relationships), Neptune is based on graph data structures (i.e., vertices, edges, properties .... more on this later).[28] Also a Neptune graph database particularly appeals to applications with *highly connected* datasets and is accessed via well-known *graph query languages* (GQLs), such as Gremlin and SPARQL (**S**PARQL **P**rotocol **A**nd **R**DF **Q**uery **L**anguage).[29]

Gremlin, developed by the Apache TinkerPop™ open source project, provides a graph data query language as well as a virtual machine -- akin to a Java Virtual Machine (JVM) -- and is suitable for both OLTP and OLAP applications.

SPARQL is a semantic query language for accessing data stored in RDF (Resource Description Framework) format.[30] SPARQL is defined as a formal specification among the family of RDF recommendations from the W3C standards organization (World Wide Web Consortium). SPARQL supports a query syntax for traversing and navigating data in a graph database.

Regarding *highly connected* datasets, graph database engines and graph query languages are well-matched for applications that require navigating *complex network relationships* and also demand rapid results from queries. Complex network relationships typically include numerous, intersecting, many-to-many entity relationships such as those found in applications known as *recommender engines*.[6] Examples of such complex, many-to-many relationships include combinations of the following:

- A person can friend (or follow, or professionally connect to) many people … and each person can be friended (or followed or connected) by many other persons (e.g., social media for recommending friends, professional connections, persons to follow … as respectively for Facebook, LinkedIn, and Twitter).

- A person purchases many products or services … each product or service can be purchased by many people. For online retail recommending …
    - Products or services
    - Similar products or services
    - Accessories of products purchased at the same time
    - Products or services purchased by people who are friends, connections or live nearby

- A person visited many places (each with many attractions, hotels, restaurants, etc.) and each place has been visited by many people, friends, connections, neighbors -- for online travel recommendations.

Basic vocabulary of graph databases includes concepts such as *vertices* (aka *nodes*), *edges*, and *properties* as the basis for describing, storing and navigating data. These concepts are comparable to fundamental RDBMS concepts (i.e., table rows, foreign keys, columns) and data modeling concepts (i.e. entities, relationships, attributes):

- ***Vertices (aka Nodes)***: Vertices or nodes are analogous to *rows* in a relational table and to *entity occurrences* using data modeling parlance. Vertices are assigned to one or more *labels* -- comparable to a table name and table synonyms in a relational database -- for distinguishing node types (i.e., entities such as person, employee, location, product, asset, etc.).

- ***Edges***: Edges express an association or relationship between vertices (i.e., an edge is a connection that has a start node and an end node). In the RDBMS world, *relationships* are implemented as *foreign keys* and described as one-to-many relationships, many-to-many relationships, or one-to-one relationships in data models and data modeling tools.

- ***Properties***: Properties are comparable to RDBMS table *columns* or *attributes* of entities in a data model. Properties are stored as a collection of *key-value pairs* (aka *attribute–value pairs*) where the *key* identifies (i.e., names or describes) the property and the *value* is the data content of the property (e.g., where *key* = 'job_title' and *value* = 'sales person').[31] The surprise is that properties apply to edges as well as vertices. Many key-value pairs can be defined for vertices and edges. Edges having properties is again reminiscent of the previously noted UML concept of an *association class* where properties provide additional information about the edge and its association between its start and end vertices.[32]

Graph databases supporting these basic graph model concepts (i.e., vertices, edges, labels and properties) are called labelled property graphs (LPGs).[33] LPG concepts are comparable to relational vocabulary but the emphasis is somewhat different. The paradigm shift for GQLs is the prominence placed on edges. GQLs elevate edge relationships to a higher stature. Edges are explicitly named (i.e., labelled) components, rather than being subtly inferred (as in the relational world) by foreign keys, creating indexes on foreign keys, and table joins for referencing and comparing values of foreign keys. The GQL focus is storing, searching and navigating edge interconnections between vertices. For many complex OLTP and OLAP applications, edges enable a direct linkage for more succinctly, intuitively, and efficiently traversing the linkages between vertices.

Gremlin Introduction

Unsurprisingly, the Gremlin query language syntax substantially differs from SQL. Specifically, Gremlin queries can concatenate a string of many retrieval operations -- referred to as graph traversal steps -- for successively navigating between vertices via edges.[34] For example, a Gremlin query can consist of a sequence of separate steps for traversing from vertex 1 (via a named edge) to vertex 2 -- and then concatenating additional steps for navigating from vertex 2 (via a differently named edge) to vertex 3, etc. The output of a step feeds into the subsequent step as its input. A single Gremlin query statement can traverse many vertices of different labels via different edge types with filters using properties of vertices or edges along the path.

Using Gremlin console commands, a brief summary of Gremlin's rudimentary graph traversal steps for **C**reating, **R**eading, **U**pdating and **D**eleting graph data (i.e., fundamental CRUD data operations) -- in contrast to corresponding SQL statements -- is included as follows:

Gremlin Graph Traversal Steps -- Basics

- ***addV*** (Add Vertex): The Add Vertex step is analogous to a SQL INSERT statement for inserting a row into an RDBMS table. The *addV* step adds a vertex to a graph and assigns one or more *labels* indicating the associated type or types to which the vertex belongs. For example, multiple labels could connote synonyms or sub-type inheritance. Properties and property values are added to a vertex by chaining one or more ***property*** *(Add Property)* steps to an *addV* step. The following statement is a simple example for adding a vertex with an ID, name and date-of-birth properties:

  g.**addV**('person').**property**(id, 'PER-0001').
  property('name','Random A. Person').**property**('dob', '03/03/1995')

  The following statement adds a person who is also an employee. Note that this includes two labels (perhaps connoting that an **employee** is a sub-type of **person**) with an additional property for the employee's job title.

  g.**addV**("person::employee").**property**(id, 'PER-0002').
  property('name', 'Random B. Person').
  property('dob', '04/04/1990').**property**('title','sales person')

- ***addE*** (Add Edge): The Add Edge step establishes a linkage between a starting vertex and an ending vertex. The *addE* step assigns a label (e.g., friend, lives near, knows, purchased, etc.) for distinguishing different types of edges that can be established between vertices. The closest analogy to the relational world -- as previously noted -- is the value of a column defined as foreign key implying a linkage to a row of another table (or the same table). The following statement is a simple example for adding an edge from a starting vertex to an ending vertex with a label "friend" indicating that a person is a friend of another person, and also includes a property for indicating the strength of the friendship:

g.**addE**('friend').from(g.V('PER-0001')).to(g.V('PER-0002')).
property('weight', 1.0)

- **Property** (Add Property): As noted in previous examples for the *addV* and *addE* steps, one or more properties can be defined for both vertices and edges. Different vertices or edges having the same label can flexibly include varying sets of properties -- unlike the structured relational approach whereby a CREATE TABLE statement rigidly specifies columns that apply to all rows of a table. There is no analog in Gremlin to a CREATE TABLE statement. The property step can be used for conveniently adding properties to new or existing vertices and edges, and is also used for updating or appending values to previously defined properties.

- **Has:** The *Has* step is used for filtering and is roughly analogous to a SQL SELECT statement for sifting rows from one or more RDBMS tables. But the *Has* step applies to both vertices and edges. It is normally used for selecting vertices and edges based on property filters, somewhat comparable to using a WHERE clause in a SQL SELECT statement. There are many variations for the *Has* step, several examples of which are included as follows:

  - **Has (key, value)**: This option specifies a key and value. It selects vertices or edges whose property value equals the specified value.
  - **Has (label, key, value)**: This option specifies a label, key and value. It selects the vertices or edges having the specified label, property and value.
  - **hasLabel (label ...)**: This option specifies one or more labels and selects the vertices or edges having at least one of the specified labels.
  - **Has (key, predicate)**: This option specifies a key and a predicate (e.g., a comparison operator and value) for comparing the property value with the predicate. Examples of comparison operators include "eq" (equal to), "neq" (not equal to), "lt" (less than), "gt" (greater than), etc. This option selects vertices or edges for which the value of the specified property satisfies the predicate. A simple example of this option combined with the hasLabel option for counting employees who are sales people is included as follows:

    g.V().**hasLabel**('employee').**has**('title',**eq**('sales person')).count()

- **Drop:** The drop step -- roughly analogous to a SQL DELETE statement for removing rows from an RDBMS table -- can be used for deleting vertices, edges, or simply removing properties of vertices or edges. A simple example for removing the subset of 'sales person' vertices is included as follows:

g.V().**has**('title','sales person').**drop**()

The previous descriptions and examples of these basic steps represent the tip of the Gremlin iceberg (i.e., a small subset of the overall comprehensive assortment of graph traversal steps supported by Gremlin). These elementary steps are included here simply to provide a glimpse at Gremlin's fundamental language concepts. A robust collection of many more steps and options are supported by Gremlin for allowing complex data transformations, filtering objects, calculating statistics, etc. Fortunately, abundant documentation describing Gremlin traversal steps (and providing copious examples with subtle nuances) is readily available online.[35]

SPARQL Introduction

Unlike Gremlin, SPARQL as previously noted is targeted for data stored in RDF format and SPARQL's syntax and semantics for querying and updating RDF data is more closely aligned with SQL. The RDF graph model (aka a triple store) stores linked data in the form of data *triples* derived from statements consisting of a *subject + predicate + object*. An RDF triple store is considered to be another type of graph database.[36] In an RDF graph, the subject and the object become vertices connected by an edge -- the predicate is an edge:

Triple Statement ➔ Subject (Vertex) + Predicate (Edge) + Object (Vertex)

Each triple statement links a subject (i.e., a vertex, typically a discrete entity occurrence such as an individual person, location, product, asset, etc.) via a predicate (edge) to another object (a vertex). In effect, the predicate names the relationship between the subject and object. The object is a discrete entity or can also be a simple literal value (e.g., a character string, date, integer, boolean, etc.) which also is stored as a vertex. Except for literals, objects in a triple statement can be the subject or object in other triple statements, thereby enabling graph traversal from subjects to objects as other subjects. Objects as literal values in essence are property values where the predicate is tantamount to being a property name as well as an edge (more about this later).

Examples of RDF triple statements, corresponding to examples of Gremlin graph traversal steps from in the previous section, are provided as follows:

- **Gremlin Statement** ...

  g.**addV**('person').**property**(id, 'PER-0001').
  property('name','Random A. Person').**property**('dob', '03/03/1995')

  Corresponding RDF Triple Statements ...

  - PER-0001 is a person
  - PER-0001 has name 'Random A. Person'
  - PER-0001 was born on '03/03/1995'

  ... Where "PER-0001" is a Subject ... "is a", "has name", "was born on" are Predicates ... "person" is an Object ... 'Random A. Person' and '03/03/1995' are literal Objects.

- **Gremlin Statement** …

g.**addV**("person::employee").**property**(id, 'PER-0002').
property('name', 'Random B. Person').
property('dob','04/04/1990').**property**('title','sales person')

Corresponding RDF Triple Statements …

- PER-0002 is an employee
- employee is a kind of person
- PER-0002 has name 'Random B. Person'
- PER-0002 was born on '04/04/1990'
- PER-0002 has job title 'sales person'

… Where "PER-0002" is a Subject, "employee" is both an Object and Subject (i.e., in 2 separate statements) … "person" is an Object … "is an", "is a kind of", "has name", "was born on", "has job title" are Predicates … 'Random B. Person', '04/04/1990', and 'sales person' are literal Objects.

Unlike Gremlin's labelled property graphs where vertices and edges can each have many properties, RDF subjects and predicates do not have properties directly associated with them. In RDF, each property of a subject/vertex is implemented as a predicate/edge linking to an object as a literal value. As noted above, the predicate serves not only as an edge but is also analogous to a property name. So, in the previous triple statement ...

PER-0002 has job title 'sales person' …

the predicate "has job title" connects PER-0002 (subject) to the literal value 'sales person' (object = a character string) and, in essence, also suggests a property name (i.e., job_title). Each object, therefore, represents either some discrete entity occurrence (i.e., an individual person, place, product, event, thing, concept, etc.) or the value of some property of the subject. So … yes … property values can indeed be linked to a subject but, unlike an LPG, this results in the creation of another vertex and edge in the RDF graph. Clearly for implementing properties of subjects, an RDF graph is likely to contain many more vertices than those that would be required for an LPG.

But what about adding a property to an RDF predicate? Analogous to adding a property to an edge in an LPG, is it possible -- in some way -- to add information about an RDF predicate? Yes … but, sadly, the method for doing such is uncomfortably convoluted!

The RDF specification provides a technique called RDF Reification for adding information about an entire triple statement (e.g., the date the statement was made, the reason for the statement, the individual who made the statement, a rating or weight of the association, etc.).[37] In essence, adding information about a triple statement itself is tantamount to adding information about the predicate. In a nutshell -- without the gory details -- reification adds four additional triple statements (aka the "reification quad") about the original statement itself. These additional statements amount to the inclusion of *metadata* within the RDF graph for describing the original statement. So, in this context, the RDF graph not only contains application data (i.e., data about the problem domain), but now unfortunately also contains metadata (i.e., data about the data -- data about the statements themselves). This reification quad, therefore, sadly introduces substantially more vertices and edges than would be required for an LPG.

More specifically, the four additional statements, the reification quad, are summarized as follows:

- A new triple statement defining the original statement itself:
    - Subject = A unique identifier for the statement itself + …
    - Predicate = "type" + …
    - Object = RDF Statement
- Three additional statements for declaring its subject, predicate, and object.

Along with these four additional statements added to the RDF graph, yet another final statement can now be created for achieving the ultimate goal of adding a property value for the original statement, i.e. for a linking another object or literal to the unique identifier for the original statement. Such complexity, in terms of the need for introducing additional vertices and edges -- for simply adding a property to a predicate -- is seemingly breath-taking in comparison to adding a property to an edge in an LPG. But that's the bad news … Here perhaps is now the good news!

The SPARQL commands for creating, reading, updating, and deleting triple statements (CRUD) as previously indicated are somewhat comparable to SQL (i.e., SQL basic CRUD DML commands: INSERT, SELECT, UPDATE, DELETE). Data professionals (i.e., DBAs, database designers, applications developers, etc.) conversant in SQL are frequently quite comfortable if the need for transitioning to SPARQL arises. Several rudimentary SPARQL commands in comparison to SQL are briefly introduced as follows:

- **SELECT**: Comparable to the SQL SELECT statement for retrieving columns of rows from one or more relational tables, the SPARQL SELECT command queries and retrieves properties from one or more triples. Many key words are common to the SPARQL SELECT and SQL SELECT commands with comparable semantics (e.g., FROM, WHERE, GROUP BY, HAVING, UNION, DISTINCT, FILTER and many aggregate functions such as SUM, AVG, MIN, MAX, COUNT, etc.). There are of course many subtle differences. For example, the optional WHERE clause specifies search criteria in the form of one or more *triple patterns*. A triple pattern is simply a triple statement where any of its components -- subject, predicate or object -- can be substituted with a variable for matching any vertex. Selected variables are returned in the query result. A FILTER key word can also be used to further constrain variable values. For example, a simple search of employees with job title "sales person" is provided as follows:

```
SELECT ?name
WHERE {
    ?employee prop:name ?name
    ?employee prop:jobtitle "sales person"
}
```

SPARQL variable names begin with a "?", hence the variables ?employee, ?name, and ?jobtitle in the previous example.

- **INSERT DATA**: Comparable to the SQL INSERT statement for adding rows to a relational table, the SPARQL INSERT DATA command adds one or more triples to a specified or default RDF graph.

- **INSERT WHERE**: Based on search criteria specified in a WHERE clause, the INSERT WHERE command is a more powerful option for inserting one or more triples into an RDF graph. In this case, existing triples matching the specified search criteria are used as the basis for inserting new triples. Search results from the WHERE clause are combined with specified triple patterns for inserting new triples bound to the search results. A simple example is the need for inserting new triples for adding properties to subjects matched in the WHERE clause. The INSERT WHERE command can also be used for copying triples based on search criteria from a source graph to another graph.

- **DELETE DATA**: Comparable to the SQL DELETE statement for removing rows from a relational table, the SPARQL DELETE DATA command removes one or more specified triples from a specified or default RDF graph.

- **DELETE WHERE**: Based on search criteria specified in a WHERE clause, the DELETE WHERE command is a more powerful option for removing one or more triples from an RDF graph. In a manner similar to the INSERT WHERE command, triples matching the specified search criteria are used as the basis for removing existing triples. Search results from the WHERE clause are combined with one or more specified triple patterns for deleting existing triples bound to the search results. A simple example is the need for removing triples for properties predicated to subjects matched in the WHERE clause.

- **DELETE/INSERT WHERE**: SQL is endowed with an UPDATE statement for directly modifying the content of existing rows in a relational table. Unlike SQL, SPARQL does not provide an analogous command for directly updating the content of existing triples. Instead, SPARQL provides the DELETE/INSERT statement -- which includes a WHERE clause -- for first removing existing triples and then inserting corresponding triples with revised information. Search results from the WHERE clause are combined with specified triple patterns for deleting existing triples and separate triple patterns for specifying updates.

Like the Gremlin introduction of the previous section, this brief introduction to SPARQL commands also represents the tip of the SPARQL iceberg (i.e., a small subset of the overall comprehensive assortment of key words and options supported by SPARQL). The command descriptions included herein simply provide a glimpse at SPARQL's fundamental language concepts. Fortunately, abundant documentation describing SPARQL commands, examples, key words, options, and complex usage (e.g., joins between triples, outer joins, subqueries, etc.) is also readily available online.[38]

Neptune Clusters

A Neptune cluster is very similar to an Aurora cluster. There are, however, some noteworthy differences that justify distinguishing Neptune clusters from Aurora clusters by establishing a new entity, the NEPTUNE CLUSTER entity. Figure 4.1 introduces entities and relationships focusing on Neptune cluster fundamental concepts.

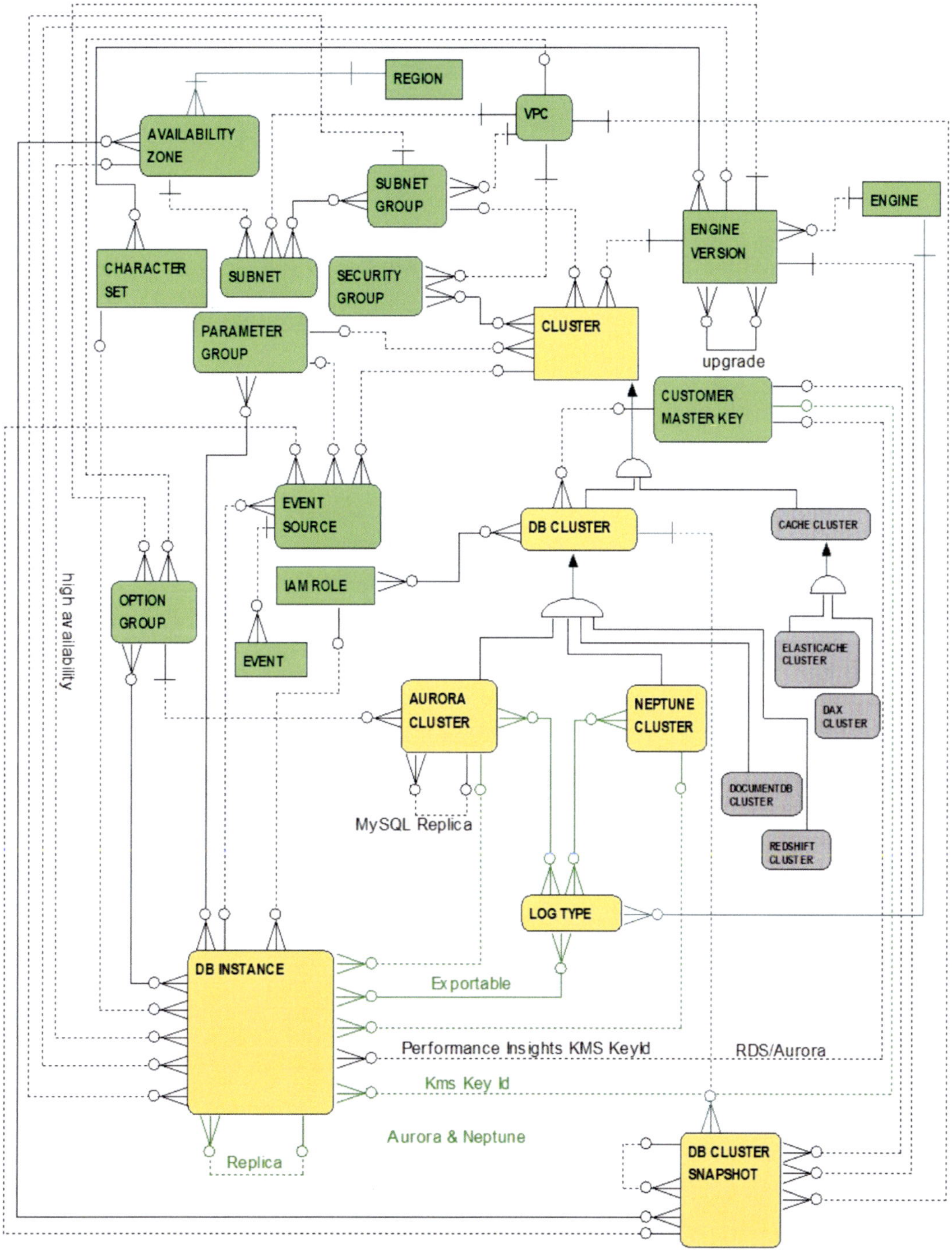

Figure 4.1 Entities and Relationships … Neptune Fundamentals

Figure 4.1 highlights the NEPTUNE CLUSTER entity and reaffirms the explanation of the CLUSTER entity and its sub-type hierarchy in the previous chapter. Similar to Aurora, Neptune also enables the provisioning of a cluster of DB instances, consisting of a primary DB instance and as many as 15 read replica DB instances. The NEPTUNE CLUSTER entity -- at the same level as the AURORA CLUSTER, DOCUMENTDB CLUSTER, and REDSHIFT CLUSTER entities -- is a sub-type of the DB CLUSTER entity. Each Neptune cluster is *a kind of* DB cluster and, therefore, directly inherits attributes and relationships from the DB CLUSTER entity. Indirectly, it also enjoys inherited attributes and relationships from the CLUSTER entity.

More specifically, as a sub-type of the DB CLUSTER entity, a Neptune cluster (i.e., the graph data maintained by all of its cluster DB instances) is optionally encrypted using a KMS customer master key, and benefits from one or more IAM roles enabling Neptune to access other AWS services on behalf of the Neptune cluster. From Figure 4.1 and as mentioned in the previous chapter, recall that a Neptune cluster -- as *a kind of* DB cluster -- can be backed up by one or more DB cluster snapshots.

Common attributes inherited by the NEPTUNE CLUSTER entity from the DB CLUSTER entity include the following sample attributes noted as well in the previous chapter:

- *backup retention period:* The number of days that automatic backups are saved prior to being purged.
- *database name:* Identifies the name of the DB cluster's initial database.
- *master username:* The master username for logging into the cluster's DB instances.
- *storage encrypted:* Indicates whether or not the cluster's data is encrypted.

Indirectly inherited from the CLUSTER entity, a Neptune cluster can serve as a source for event notifications and is associated with many security groups, a single subnet group, parameter group, and an engine version. This means that a cluster of Neptune DB instances is blessed with …

- A set of security groups for controlling access to Neptune cluster instances.
- A subnet group within whose subnets Neptune cluster instances can be launched.
- A single parameter group specifying parameters for configuring all Neptune cluster instances.
- A Neptune engine version for all cluster instances.

Common attributes indirectly inherited by the NEPTUNE CLUSTER entity from the CLUSTER entity include the sample attributes noted in the previous chapter:

- *cluster identifier:* A user-supplied cluster identifier.
- *cluster create time:* The time when the cluster was created.
- *status:* Indicates the current state of the cluster.
- *endpoint:* Specifies the cluster's access information (i.e., DNS host name and port number) enabling applications to connect to the cluster.
- *preferred maintenance window:* Identifies the time period when maintenance activities (e.g., system upgrades) are preferred.

The one-to-many relationship from NEPTUNE CLUSTER entity to the DB INSTANCE entity indicates that each Neptune cluster is composed of one or more DB instances (i.e., all DB instances running the same Neptune engine version). One DB instance is primary and the remaining servers are read replicas. For example, via the Neptune console, when creating a Neptune cluster, a primary DB instance is automatically created for reading, writing and modifying data to the cluster volume. One or more replica DB instances -- possibly in separate availability zones for achieving high availability -- can be added to a Neptune cluster using the Neptune console, CLI, or API.

From 4.1, the optional one-to-many relationships from AURORA CLUSTER and NEPTUNE CLUSTER entities to the DB INSTANCE entity informally convey that each DB instance is exclusively either: 1) A member of an Aurora cluster, or 2) A member of a Neptune cluster, or 3) Neither -- simply a standalone RDS instance.

After a Neptune cluster is created, conventional tools and utilities (i.e., Gremlin, SPARQL, or Neptune utilities) running on client EC2 instances are used for connecting to graph databases. Examples of such utilities include the Gremlin Console, the RDF4J Console for SPARQL access,[39] and native Neptune endpoints enabling HTTP/REST queries for Gremlin and SPARQL.

Similar to Aurora, the one-to-many reflexive relationship from the DB INSTANCE entity to itself indicates that possibly many DB instances can serve as read replicas for the primary Neptune DB instance (i.e., a primary DB instance is synchronized to possibly many read replicas). Also similar to Aurora, the one-to-many relationship between CUSTOMER MASTER KEY and DB INSTANCE entities -- for encrypting DB instance data -- applies to Neptune DB instances. Also similar to Aurora, the many-to-many relationship between NEPTUNE CLUSTER and LOG TYPE entities affirms Neptune support for publishing log files to Amazon CloudWatch Logs.

In addition to previously described attributes inherited from its super-type hierarchy (i.e., foreign keys and other attributes inherited from the CLUSTER and DB CLUSTER entities), noteworthy examples of attributes pertinent to the NEPTUNE CLUSTER entity include:

- *multi AZ*: Indicates high availability in that DB instances are provisioned in more than one availability zone.
- *preferred backup window:* If automatic backups are enabled, this indicates the time range (i.e., a user-defined daily time window) when automated backups are created.
- *earliest restorable time:* For restoring a cluster to a specific point-in-time, the earliest possible time in the past to the cluster can be recovered.
- *latest restorable time*: For restoring a cluster to a specific point-in-time, the latest possible time in the past the cluster can be recovered.
- *allocated storage*: Always set to 1 for Neptune indicating that Neptune storage size is not fixed but automatically scales as necessary.

Recall from the previous chapter that these immediately preceding attributes also apply to Aurora clusters. But these attributes are not applicable (as will be explored in subsequent chapters, e.g., Chapter 5 for DocumentDB) to other sub-types of the DB CLUSTER entity. These attributes, therefore, cannot be elevated to the DB CLUSTER entity thereby enabling inheritance by all of its sub-types.

Aurora and Neptune cluster similarities suggest that perhaps Aurora and Neptune clusters might be generalized to yet another higher level of abstraction for hosting their common relationships and attributes (i.e., that the AURORA CLUSTER and NEPTUNE CLUSTER entities become sub-types of a new super-type entity beneath the DB CLUSTER super-type). The author, however, chose not to introduce this additional layer of complexity and defers exploring such (i.e., modifying Figure 4.1 to depict the suggested super-type) as an exercise for curious data modeling students. Recall from the explanation of a conceptual data model (CDM) and a more detailed logical data model (LDM) in the Introduction, that such an exercise might arguably be more appropriate and justifiable when evolving to an LDM from a CDM.

Neptune vs. Aurora Cluster Differences

Clearly from the previous section, a Neptune cluster is similar to an Aurora cluster. Aurora clusters, however, exhibit several noteworthy curiosities in contrast with Neptune clusters which vindicate establishing the NEPTUNE CLUSTER entity as a justifiably distinct sub-type of the DB CLUSTER entity. Figure 4.2 highlights Aurora vs. Neptune major conceptual differences.

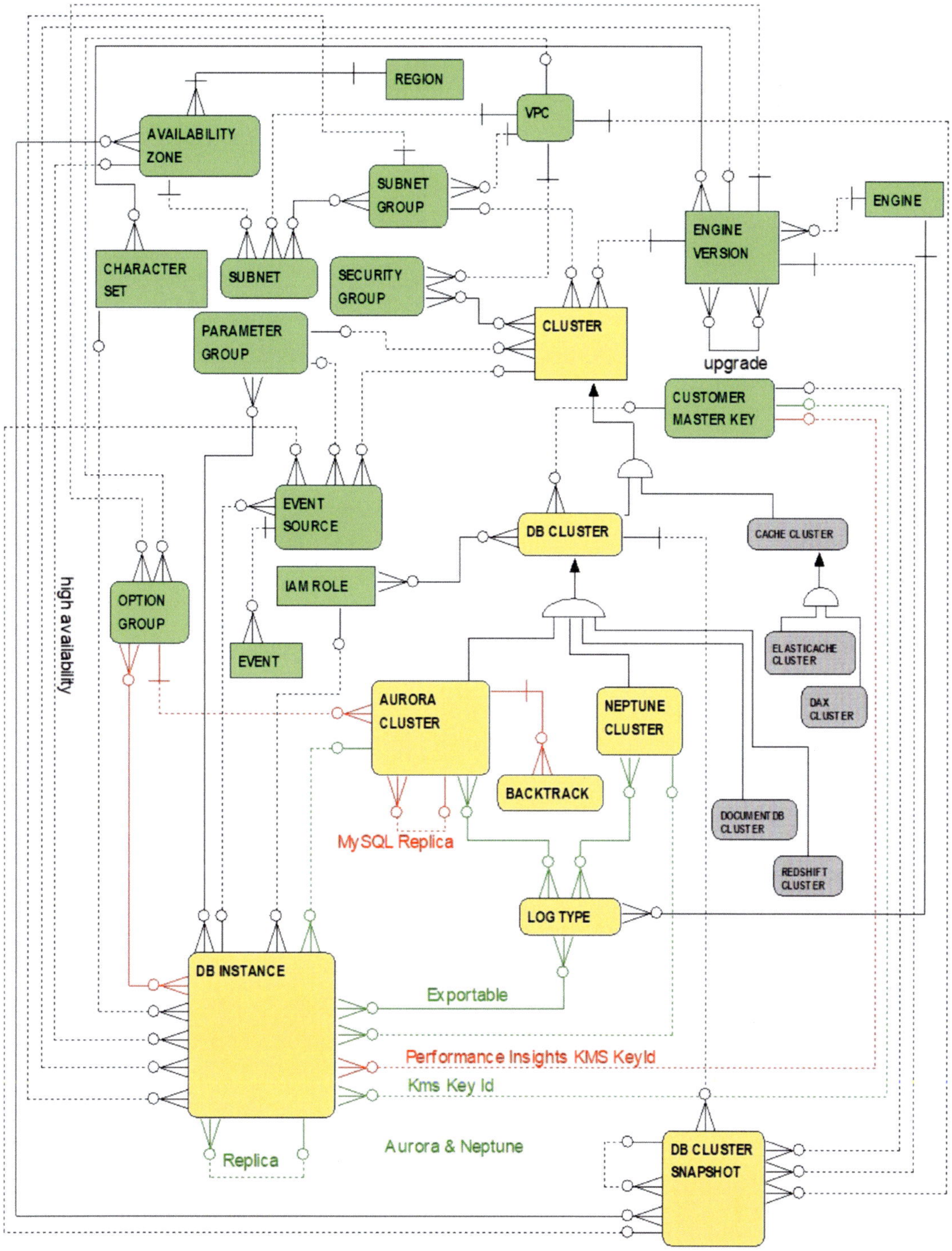

Figure 4.2 Entities and Relationships Highlighting Aurora vs. Neptune

From Figure 4.2, the following similarities (i.e., relationships highlighted in green), re-emphasized from the previous section for the reader's convenience, are highlighted as follows:

- Each Aurora and Neptune cluster is *a kind of* DB cluster (i.e., the AURORA CLUSTER and NEPTUNE CLUSTER entities are sub-types of the DB CLUSTER entity) and as such enjoy all of the relationships and attributes directly inherited from their parent DB CLUSTER entity and indirectly inherited from their grandparent CLUSTER entity.

- Both Aurora and Neptune clusters each have a one-to-many relationship to the DB INSTANCE entity and share similar concepts and vocabulary. Each cluster is composed of one or more DB instances. One DB instance is primary and the remaining servers are read replicas.
- The reflexive relationship from the DB INSTANCE entity to itself -- for read replicas -- applies to both Aurora and Neptune DB instances. This indicates that possibly many DB instances that can serve as read replicas for a primary DB instance (i.e., a primary DB instance is synchronized to possibly many read replicas).
- Many-to-many relationships to the LOG TYPE entity highlight both Aurora and Neptune support for publishing log files to Amazon CloudWatch Logs.
- The one-to-many relationship between CUSTOMER MASTER KEY and DB INSTANCE entities -- for encrypting DB instance data -- applies to both Aurora and Neptune DB instances.

Figure 4.2 also highlights AURORA CLUSTER relationships to the BACKTRACK entity, OPTION GROUP entity, and its reflexive relationship (i.e., relationships highlighted in red). The differences? Unlike Aurora …

- Neptune does not support backtracking and serverless clusters.
- Neptune does not support the concept of option groups. Specifically, the one-to-many relationship from the OPTION GROUP entity to the AURORA CLUSTER entity -- reflecting the default option group applicable to all DB instances in an Aurora cluster -- is not applicable to Neptune clusters.
- Similarly, the many-to-many relationship between the DB INSTANCE and OPTION GROUP entities -- reflecting the option groups applicable to DB instances -- is not applicable to Neptune DB instances.
- The one-to-many relationship between CUSTOMER MASTER KEY and DB INSTANCE entities -- labelled "Performance Insights KMS Key Id" for encrypting *Performance Insights* metrics data -- is not applicable to Neptune DB instances.
- Neptune does not support cross-region replication. Specifically, the one-to-many reflexive relationship from AURORA CLUSTER entity to itself (highlighting support for cross-region replication for MySQL) is not applicable to Neptune clusters.

Lack of Neptune support for backtracking and serverless clusters means that the following attributes -- conveniently repeated here from the previous chapter -- are applicable to Aurora clusters but are **not applicable** to Neptune clusters:

- *backtrack consumed change records*: The count of backtrack change records retained for the cluster.
- *backtrack window*: Amount of time in seconds for which a backtrack may be wanted, (e.g., a target backtrack window of 48 hours).
- *earliest backtrack time*: The earliest time for which a DB cluster backtrack can be requested (e.g., based on accumulated change records and available storage).
- *DB engine mode*: Indicates the engine mode (e.g., serverless, provisioned).

- *max capacity*: For a serverless cluster, the maximum capacity in terms of *Aurora capacity units* (ACUs) expressing the peak capacity (i.e. compute and memory) to which Aurora can automatically expand.

Chapter Summary

Figure 4.3 extends Figure 4.2 with additional commentary summarizing Amazon Neptune basic concepts along with Aurora similarities and differences.

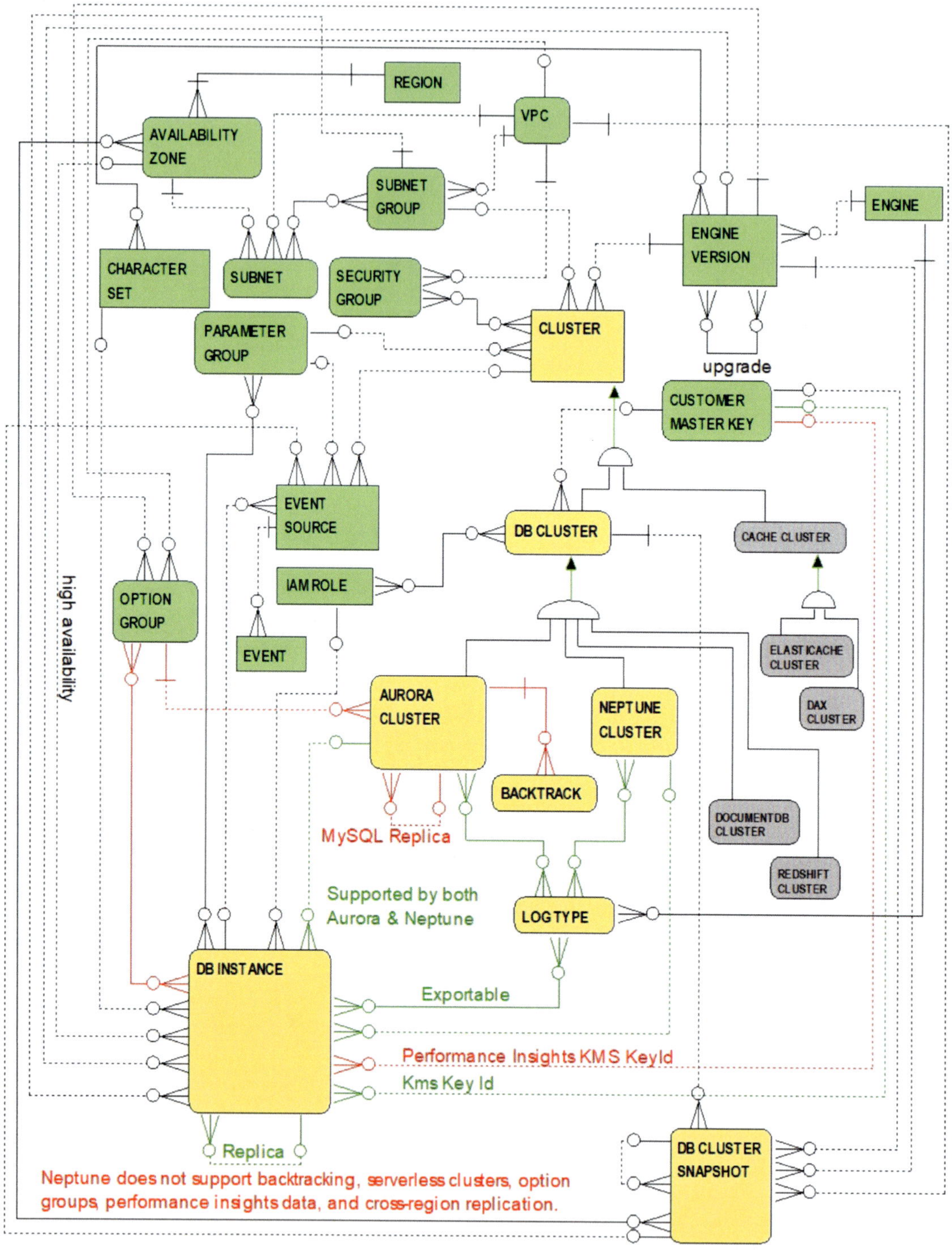

Figure 4.3 Neptune Summary with Aurora Similarities and Differences

Neptune is a *graph database* engine, a managed platform AWS database service, and (like Aurora) provides solid-state virtual storage facilities (improves database performance, reliability, and availability). Neptune also enables high availability by synchronizing database updates to read replicas and replicating data across availability zones.

A Neptune graph database (a NoSQL approach) differs from SQL-based relational engines. Neptune is based on graph data structures (i.e., vertices, edges, properties) instead of relational concepts (i.e., tables, columns, foreign keys) and accommodates apps with *highly connected* datasets using GQLs, such as Gremlin and SPARQL. GQLs are well-suited for apps navigating *complex network relationships* (numerous, many-to-many entity relationships) such as those found in contemporary *recommender engines* (e.g., friend of friend, similar products purchased, attractions visited, etc.).

Graph databases concepts include *vertices* (aka *nodes*), *edges*, and *properties* -- comparable to relational concepts (table rows, foreign keys, columns) and data modeling concepts (entities, relationships, attributes). Graph databases supporting these concepts are called labelled property graphs (LPGs). LPGs emphasize *edges* as named (i.e., labelled) components (rather than relationships inferred by RDBMS foreign keys and table joins). For complex OLTP and OLAP applications, edges enable direct linkage for efficiently traversing vertices.

Gremlin console commands -- rudimentary graph traversal steps for CRUD data operations include: *addV* (Add Vertex), *addE* (Add Edge), Property (Add Property), *Has* (for filtering, analogous to a SQL SELECT), *Drop* (analogous to a SQL DELETE).

SPARQL accesses data stored in RDF format and SPARQL's syntax for querying and updating RDF data is closer to SQL. The RDF graph model (aka a triplestore) stores linked data in the form of data *triples* derived from statements consisting of a *subject + predicate + object*. In an RDF graph, the subject and object are vertices connected by an edge. The predicate is an edge and names the relationship between the subject and object. The object is either a discrete entity or a simple literal value (e.g., a string, date, integer, etc.) which also is stored as a vertex. Except for literals, objects in a triple statement can be the subject or object in other triple statements (enabling traversal). Objects as literal values are property values where the predicate is tantamount to being a property name as well as an edge.

Analogous to adding a property to an edge in an LPG, it is also possible to add information about an RDF predicate via a technique called RDF Reification. This involves adding information about an entire triple statement (e.g., the date the statement was made, the reason for the statement, etc.). Adding information about a triple statement is tantamount to adding information about the predicate. RDF Reification requires adding metadata triple statements (the reification quad) describing the statement itself and introduces more vertices and edges than would be required for an LPG.

The SPARQL commands for creating, reading, updating, and deleting triple statements (CRUD) include: SELECT (comparable to SQL SELECT), INSERT DATA (comparable to the SQL INSERT), INSERT WHERE (based on search criteria specified in a WHERE clause), DELETE DATA (comparable to SQL DELETE), DELETE WHERE (based on search criteria specified in a WHERE clause), and DELETE/INSERT WHERE (for changing data).

Neptune, like Aurora, supports *database clustering*. A Neptune cluster consists of a primary DB instance (for read and write operations) and as many as 15 read replica DB instances (updates are automatically synchronized from the primary). Neptune replicas offload data retrieval requests. Neptune automatically switches to a read replica if the primary DB instance crashes. Also similar to Aurora, a Neptune cluster can support one or more log types indicating the types of log files that can be published (by all DB instances in the cluster) to Amazon CloudWatch Logs.

A Neptune cluster, like an Aurora cluster, is *a kind of* DB cluster which in turn is *a kind of* cluster. A Neptune cluster inherits a KMS customer master key (i.e., a CMK for data encryption), IAM roles, and cluster snapshots because it is *a kind of* DB cluster. A Neptune cluster indirectly inherits (because a DB cluster is *a kind of* cluster) many security groups, a subnet group, a parameter group, an engine version, and is a source for event notifications. After a Neptune cluster is created, conventional tools and utilities running on client EC2 instances are used for connecting to graph databases. Examples of such utilities include the Gremlin Console, the RDF4J Console for SPARQL access, and native Neptune endpoints enabling HTTP/REST queries for Gremlin and SPARQL.

Unlike Aurora, Neptune does not support backtracking, serverless clusters, option groups, *Performance Insights* metrics data, and cross-region replication.

Chapter 5 – Amazon DocumentDB

Amazon DocumentDB Introduction

Amazon DocumentDB (DocumentDB) -- like RDS, Aurora and Neptune -- is yet another managed, platform-based data service offered by AWS, and thereby streamlines many mundane database admin tasks generally performed by DBAs. Also like Aurora and Neptune, DocumentDB is equipped with solid-state virtual storage facilities which are optimized for database workloads and deliberately engineered to significantly improve database performance, reliability, and availability. DocumentDB also promotes high availability by enabling automatic failover to read replicas and data replication across availability zones.

Like Neptune, DocumentDB provides a fundamentally different alternative to SQL-based relational database approaches such as RDS, Aurora and Redshift (Redshift will be addressed next in Chapter 6). Instead, DocumentDB is an example of a *document database* (aka "document-oriented database" or "document store") and boasts compatibility with MongoDB which is also categorized as a NoSQL database service.

NoSQL, in this case, essentially means that DocumentDB provides a non-relational approach to managing semi-structured data based on key-value pairs and document storage. In particular, DocumentDB stores semi-structured data as documents in its document database. Documents are *self-describing* basically because they are composed of key-value pairs defined in a JSON-like format (i.e., BSON, a binary representation of JSON).[40] DocumentDB's compatibility with MongoDB suggests that applications developed using widely available MongoDB APIs, drivers, utilities and tools -- supported by many contemporary and popular programming languages -- will run compatibly in DocumentDB.

A document database -- in contrast to a relational database -- is very appealing to certain applications programmers who are comfortable with using the format of a document for accessing and changing data in a program. Application data can then be saved to a database in the same intuitive document format. For example, applications developers who advocate an agile software development approach (i.e., routinely anticipating frequent database design changes, in contrast to a more structured waterfall methodology) might prefer the flexibility offered by a document database.[41,42] This perspective differs from well-known SQL-based relational approaches, such as those database engines supported by RDS, for manipulating data stored in tables, rows and columns.

A brief introduction to DocumentDB API vocabulary in contrast to SQL is summarized as follows:[43]

- ***Collection***: A DocumentDB collection is a set of documents and corresponds to a table in an RDBMS database engine.
- ***Document***: A document stores semi-structured data and uses key-value pairs (fields with values) to self-describe the structure of the data (i.e., the data schema) within the document. Each JSON-like formatted document, therefore, corresponds

to a row in a relational table. Differently formatted documents can be maintained in a document database.

- ***Field***: Each document is composed of one or more fields each defined as a key-value pair. A field, therefore, somewhat corresponds to a column of a row in a relational table.
- ***Embedded Document***: Key-values pairs can be nested in a document thereby enabling complex documents which contain other *embedded* documents. An embedded document groups fields at a lower level.
- ***Object Id***: An object id is a required field serving as a document's unique identifier and, therefore, corresponds to the concept of a primary key for a row in a relational table. The object id can be specified when a document is created or, otherwise, is automatically generated.
- ***Array***: An array is a field mapped to a list of multiple values, such as an array of numbers or an array of character strings as values.

An example of a simple document is provided as follows:

```
{
"SSN": "123-45-6789",
“EmployeeID”: “PER-0001”,
"Name": "Random A. Person",
"DOB": "1990-01-01",
“Jobtitle”: “sales person”,
"Street": "1000 Any Street",
"City": "Any Town",
"State-Province": "NY",
"Country": "USA"
}
```

A subset of MongoDB CRUD operations (i.e., for creating, reading, updating, and deleting documents) is summarized as follows:

- ***insertOne***: Analogous to a SQL INSERT statement, this operation creates a new document and inserts it into a specified collection. This operation creates the specified collection if it does not already exist.
- ***insertMany***: Inserts an array of multiple documents into a specified collection. This operation also creates the specified collection if it does not already exist.
- ***find***: Analogous to a SQL SELECT statement, this operation retrieves one or more documents from a specified collection. Various options are possible for specifying search criteria and specific fields to be returned.
- ***updateOne***: Analogous to a SQL UPDATE statement, this operation updates a single document within a specified collection based on search criteria. Values for existing fields can be modified or new fields can be added to the document.
- ***updateMany***: This method updates all documents that satisfy specified search criteria for a specified collection.
- ***deleteOne***: Analogous to a SQL DELETE statement, this operation removes a single document from a specified collection based on search criteria.
- ***deleteMany***: This method removes all documents that satisfy specified search criteria from a specified collection.

Documentation for many additional MongoDB operations including usage examples and variations of the above sampled operations is readily and freely available online.[44]

DocumentDB Clusters

DocumentDB clusters are very similar to Aurora and Neptune clusters. However, a few noteworthy differences again justify differentiating DocumentDB clusters (from Aurora and Neptune clusters) by establishing a new entity, the DOCUMENTDB CLUSTER entity. Figure 5.1 introduces entities and relationships focusing on DocumentDB cluster fundamental concepts.

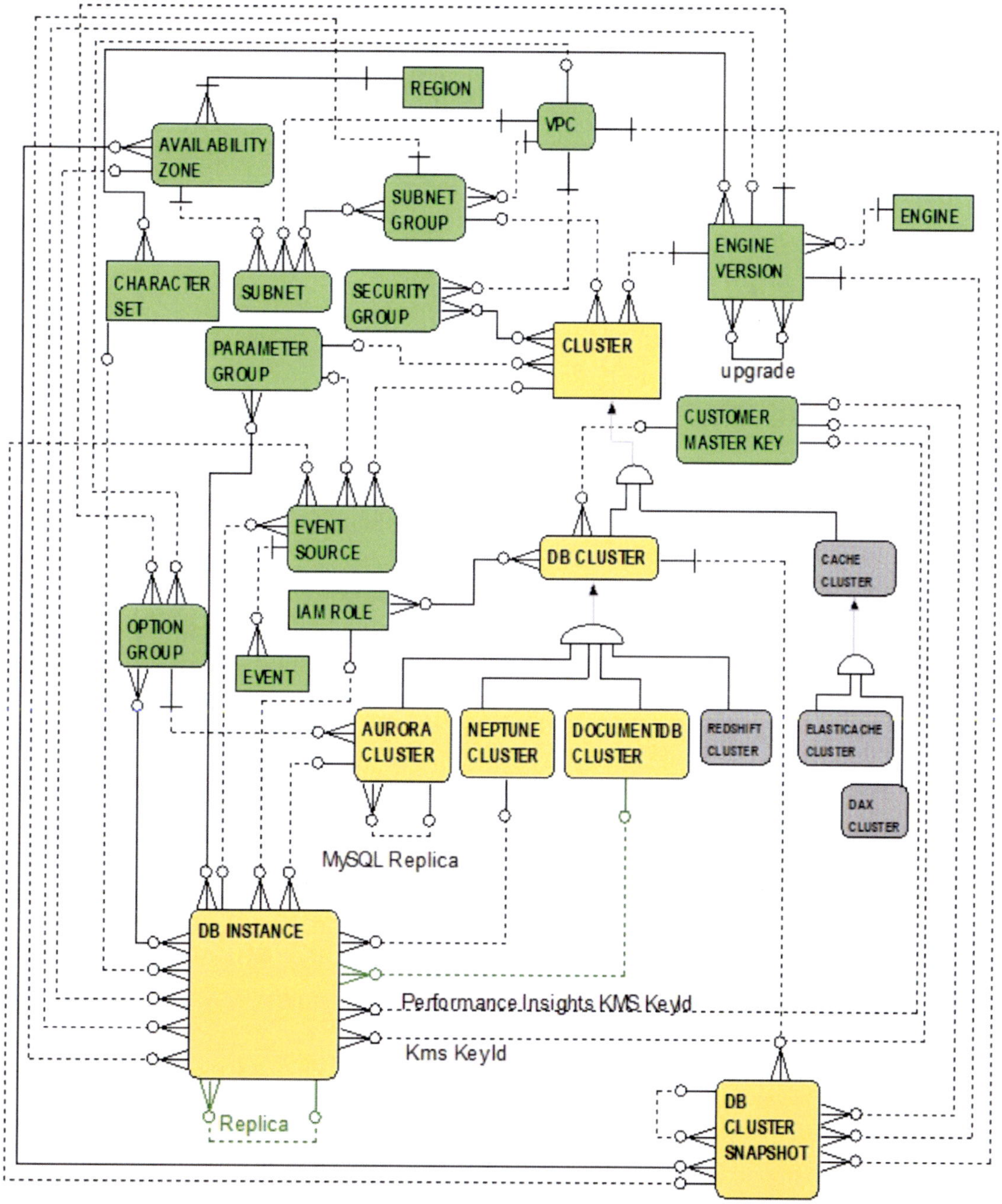

Figure 5.1 DocumentDB Entities and Relationships, Basic Concepts

Figure 5.1 highlights the DOCUMENTDB CLUSTER entity and reaffirms the explanation of the CLUSTER entity and its sub-type hierarchy -- but this time emphasizing support for DocumentDB clusters. DocumentDB -- similar to Aurora and Neptune -- also enables the provisioning of a cluster of DB instances, consisting of a primary DB instance and as many as 15 read replica DB instances. The DOCUMENTDB CLUSTER entity -- at the same level as the AURORA CLUSTER, NEPTUNE CLUSTER and REDSHIFT CLUSTER entities -- is a sub-type of the DB CLUSTER entity. Each DOCUMENTDB cluster is *a kind of* DB cluster and, therefore, directly inherits attributes and relationships from the DB CLUSTER entity. Indirectly, it also inherits attributes and relationships from the CLUSTER entity.

More specifically, as a sub-type of the DB CLUSTER entity, a DocumentDB cluster (i.e. the document data maintained by all of its cluster DB instances) is optionally encrypted using a KMS customer master key, and benefits from one or more IAM roles enabling DocumentDB to access other AWS services on behalf of the DocumentDB cluster. From Figure 5.1 and as mentioned in the previous chapter, recall that a DocumentDB cluster -- as *a kind of* DB cluster -- can be backed up by one or more DB cluster snapshots.

Common attributes inherited by the DOCUMENTDB CLUSTER entity from the DB CLUSTER entity include the following sample attributes also noted in previous chapters for Aurora and Neptune and repeated here for the reader's convenience:

- *backup retention period:* The number of days that automatic backups are saved prior to being purged.
- *database name:* Identifies the name of the DB cluster's initial database.
- *master username:* The master username for logging into the cluster's DB instances.
- *storage encrypted:* Indicates whether or not the cluster's data is encrypted.

Indirectly inherited from the CLUSTER entity, a DocumentDB cluster can serve as a source for event notifications and is associated with many security groups, a single subnet group, parameter group, and engine version. This means that a cluster of DB instances for a DocumentDB cluster has:

- A set of security groups for controlling access to DocumentDB cluster instances.
- A subnet group within whose subnets DocumentDB cluster instances can be launched.
- A single parameter group specifying parameters for configuring all cluster instances.
- The DocumentDB engine version for all cluster instances.

Common attributes indirectly inherited by the DOCUMENTDB CLUSTER entity from the CLUSTER entity include the sample attributes noted in previous chapters, repeated here as follows:

- *cluster identifier:* A user-supplied cluster identifier.
- *cluster create time:* The time when the cluster was created.
- *status:* Indicates the current state of the cluster.

- *endpoint:* Specifies the cluster's access information (i.e., DNS host name and port number) enabling applications to connect to the cluster.
- *preferred maintenance window:* Identifies the time period when maintenance activities (e.g., system upgrades) are preferred.

The one-to-many relationship from DOCUMENTDB CLUSTER entity to the DB INSTANCE entity indicates that each DocumentDB cluster is composed of one or more DB instances (i.e., all DB instances running the same DocumentDB engine version). One DB instance is primary and the remaining servers are read replicas. For example, when creating a DocumentDB cluster via the DocumentDB console, a user specifies the total number of DB instances to be launched for the cluster. This includes the primary DB instance -- automatically created for reading, writing and modifying data to the cluster volume -- and the remaining DB instances created as read replicas. By default a primary instance and two additional read replica DB instances are launched.

From 5.1, the three optional one-to-many relationships from AURORA CLUSTER, NEPTUNE CLUSTER and DOCUMENTDB CLUSTER entities to the DB INSTANCE entity are meant to convey that each DB instance is exclusively either: 1) a member of an Aurora cluster, or 2) a member of a Neptune cluster, or 3) a member of a DocumentDB cluster, or 4) neither ... simply a standalone RDS instance not participating in a cluster.

One or more replica DB instances -- possibly in separate availability zones for achieving high availability -- can be added to a DocumentDB cluster using the DocumentDB console, CLI, or API. After a DocumentDB cluster is created, conventional MongoDB tools and utilities (e.g., the MongoDB Shell for connecting and executing CRUD commands,[45] as well as MongoDB drivers and APIs for a variety of programming languages.) running on client EC2 instances are used for developing DocumentDB applications.

As for Aurora and Neptune, the one-to-many reflexive relationship from the DB INSTANCE entity to itself also applies to DB instances for a DocumentDB cluster. This indicates that possibly many DB instances can serve as read replicas for a primary DB instance (i.e., a primary DB instance is synchronized to possibly many read replicas). From a data modeling perspective, note that for Aurora and Neptune this fact is substantiated -- in API documentation for both Aurora and Neptune -- by the presence of a *Read Replica Source DBInstance Identifier* as an attribute of the *DB Instance* data type. For a read replica DB instance, this attribute contains the ID of the primary DB instance (i.e., the replica's source DB instance) and amounts to a foreign key. This attribute, however, is omitted from the *DB Instance* data type in the DocumentDB developer guide. Nevertheless, from the standpoint of creating a conceptual data model, this concept still applies to DocumentDB and is valid for describing the notion of data replication from the primary instance to its read replicas.

In addition to previously described attributes inherited from its super-type hierarchy (i.e., foreign keys and other attributes inherited from the CLUSTER and DB CLUSTER entities), noteworthy examples of attributes pertinent to the DOCUMENTDB CLUSTER entity include:

- *multi AZ*: Indicates high availability in that DB instances are provisioned in more than one availability zone.
- *preferred backup window:* If automatic backups are enabled, this indicates the time range, a user-defined daily time window, when automated backups are created.
- *earliest restorable time:* For restoring a cluster to a specific point-in-time, the earliest possible time in the past the cluster can be recovered.
- *latest restorable time:* For restoring a cluster to a specific point-in-time, the latest possible time in the past the cluster can be recovered.

DocumentDB, Neptune and Aurora Cluster Comparison

Clearly from the previous section, DocumentDB clusters are similar to Aurora and Neptune clusters. DocumentDB clusters, however, have several noteworthy differences in contrast to Aurora and Neptune clusters which justify establishing the DOCUMENTDB CLUSTER entity as a separate sub-type of the DB CLUSTER entity. Figure 5.2 first highlights DocumentDB, Aurora and Neptune similarities.

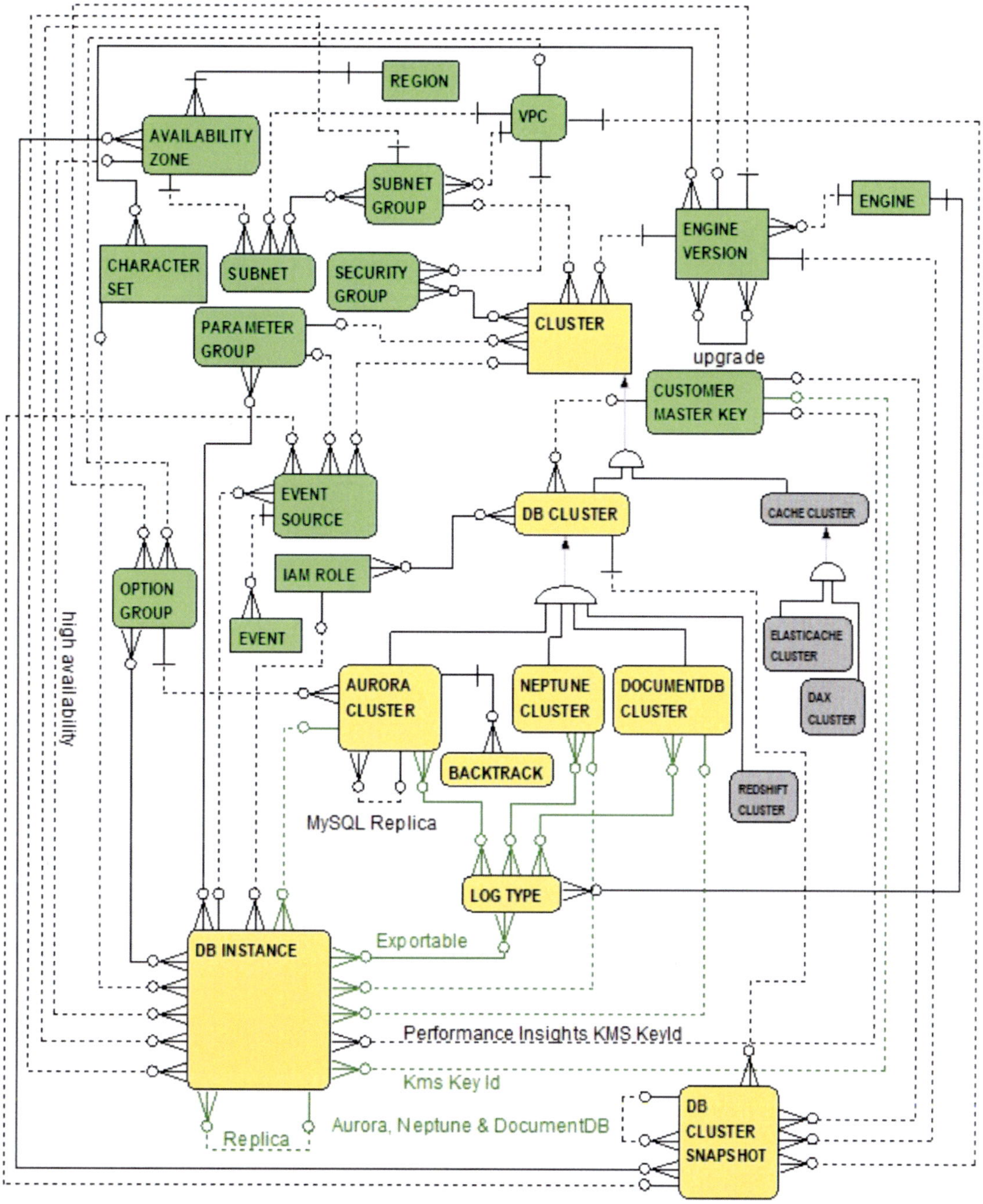

Figure 5.2 DocumentDB, Aurora and Neptune Similarities

From Figure 5.2, the following similarities are highlighted (i.e., relationships highlighted in green):

- Each DocumentDB, Aurora and Neptune cluster is *a kind of* DB cluster (i.e., the DOCUMENTDB CLUSTER, AURORA CLUSTER and NEPTUNE CLUSTER entities are sub-types of the DB CLUSTER entity) and as such enjoy all of the relationships and attributes directly inherited from their parent DB CLUSTER entity and indirectly inherited from their grandparent CLUSTER entity.
- DocumentDB, Aurora and Neptune clusters each have a one-to-many relationship to the DB INSTANCE entity and share similar concepts and vocabulary. Each

cluster is composed of one or more DB instances. One DB instance is primary and the remaining servers are read replicas.

- The one-to-many reflexive relationship from the DB INSTANCE entity to itself applies equally as well to DocumentDB, Aurora and Neptune clusters. This indicates that possibly many DB instances that can serve as read replicas for a primary DB instance (i.e., a primary DB instance is synchronized to possibly many read replicas).
- Also similar to Aurora and Neptune, the many-to-many relationship between DOCUMENTDB CLUSTER and LOG TYPE entities affirms DocumentDB support for publishing log files to Amazon CloudWatch Logs.

Figure 5.3 highlights DocumentDB, Aurora and Neptune differences.

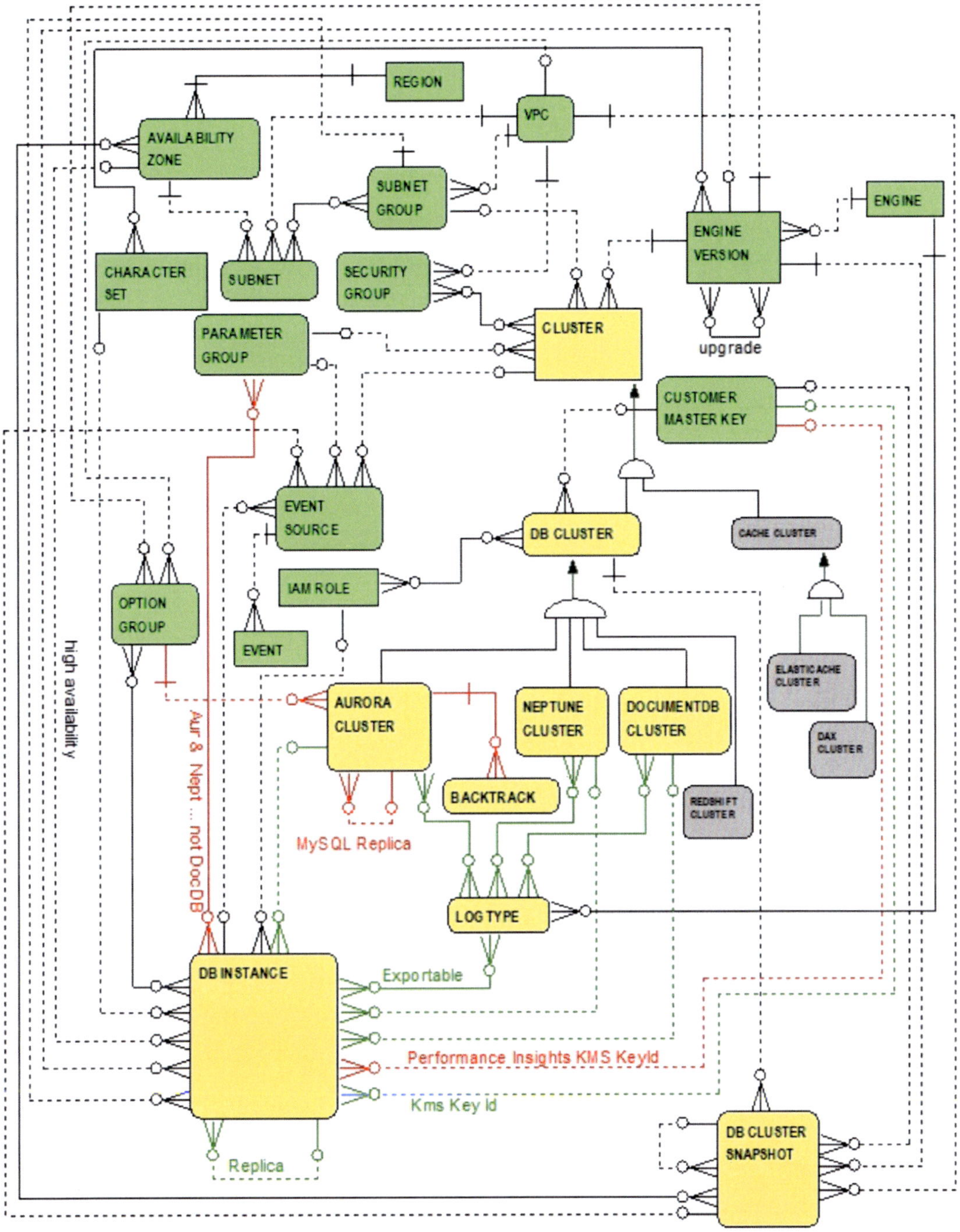

Figure 5.3 DocumentDB, Aurora and Neptune Differences

From Figure 5.3, the following differences are highlighted (i.e., relationships highlighted in red):

- Unlike Aurora, both DocumentDB and Neptune do not support backtracking and serverless clusters. Lack of DocumentDB and Neptune support for backtracking and serverless clusters means that the following attributes -- conveniently repeated here from the previous chapter -- are applicable to Aurora clusters but are **not applicable** to DocumentDB and Neptune clusters:

   - *backtrack consumed change records*: The count of backtrack change records retained for the cluster.
   - *backtrack window*: Amount of time in seconds for which a backtrack may be wanted, (e.g., a target backtrack window of 48 hours).
   - *earliest backtrack time*: The earliest time for which a DB cluster backtrack can be requested (e.g., based on accumulated change records and available storage).
   - *DB engine mode*: Indicates the engine mode (e.g., serverless, provisioned).
   - *max capacity*: For a serverless cluster, the maximum capacity in terms of *Aurora capacity units* (ACUs) expressing the peak capacity (i.e. compute and memory) to which Aurora can automatically expand.

- Unlike Aurora, both DocumentDB and Neptune do not support cross-region replication. Note the one-to-many reflexive relationship from the AURORA CLUSTER entity to itself. Comparable reflexive relationships for DOCUMENTDB CLUSTER and NEPTUNE CLUSTER entities are not applicable to DocumentDB and Neptune clusters.
- Unlike Aurora, both DocumentDB and Neptune do not support the concept of option groups. Specifically, the one-to-many relationship from the OPTION GROUP entity to the AURORA CLUSTER entity -- reflecting the default option group applicable to all DB instances in an Aurora cluster -- is not applicable to DocumentDB and Neptune clusters.
- Similarly, the many-to-many relationship between the DB INSTANCE and OPTION GROUP entities -- reflecting the option groups applicable to DB instances -- is not applicable to DocumentDB and Neptune DB instances.
- The one-to-many relationship between CUSTOMER MASTER KEY and DB INSTANCE entities -- labelled "Performance Insights KMS Key Id" for encrypting Performance Insights metrics data -- is not applicable to DocumentDB and Neptune DB instances.
- Unlike Aurora and Neptune, DocumentDB does not support parameter groups for DB instances. Figure 5.3 is appropriately annotated to reflect that the many-to-many relationship between the PARAMETER GROUP and DB INSTANCE entities is irrelevant to DocumentDB.

The astute data modeling student might infer (i.e., by noting the similarities between the DOCUMENTDB CLUSTER and NEPTUNE CLUSTER entities and their shared differences with the AURORA CLUSTER entity) that perhaps DocumentDB and Neptune clusters might be generalized to yet another higher level of abstraction for hosting their common relationships and attributes (i.e., that the DOCUMENTDB CLUSTER and NEPTUNE CLUSTER entities become sub-types of a new super-type entity beneath the DB CLUSTER super-type). Such could be done. The author, however, chose not to so. Why? Simply because abstracting these to a higher level would introduce an additional layer of complexity to the conceptual model. Exploring such (i.e., modifying Figure 5.1 to depict the suggested super-type) is deferred as an exercise for curious data modeling students. Recall from the explanation of a conceptual data model (CDM) and a more detailed logical data model (LDM) in the Introduction, that such an exercise might arguably be more appropriate and justifiable when evolving to an LDM from a CDM.

Chapter Summary

Figure 5.4 extends Figure 5.3 with additional commentary summarizing Amazon DocumentDB basic concepts along with Aurora and Neptune similarities and differences.

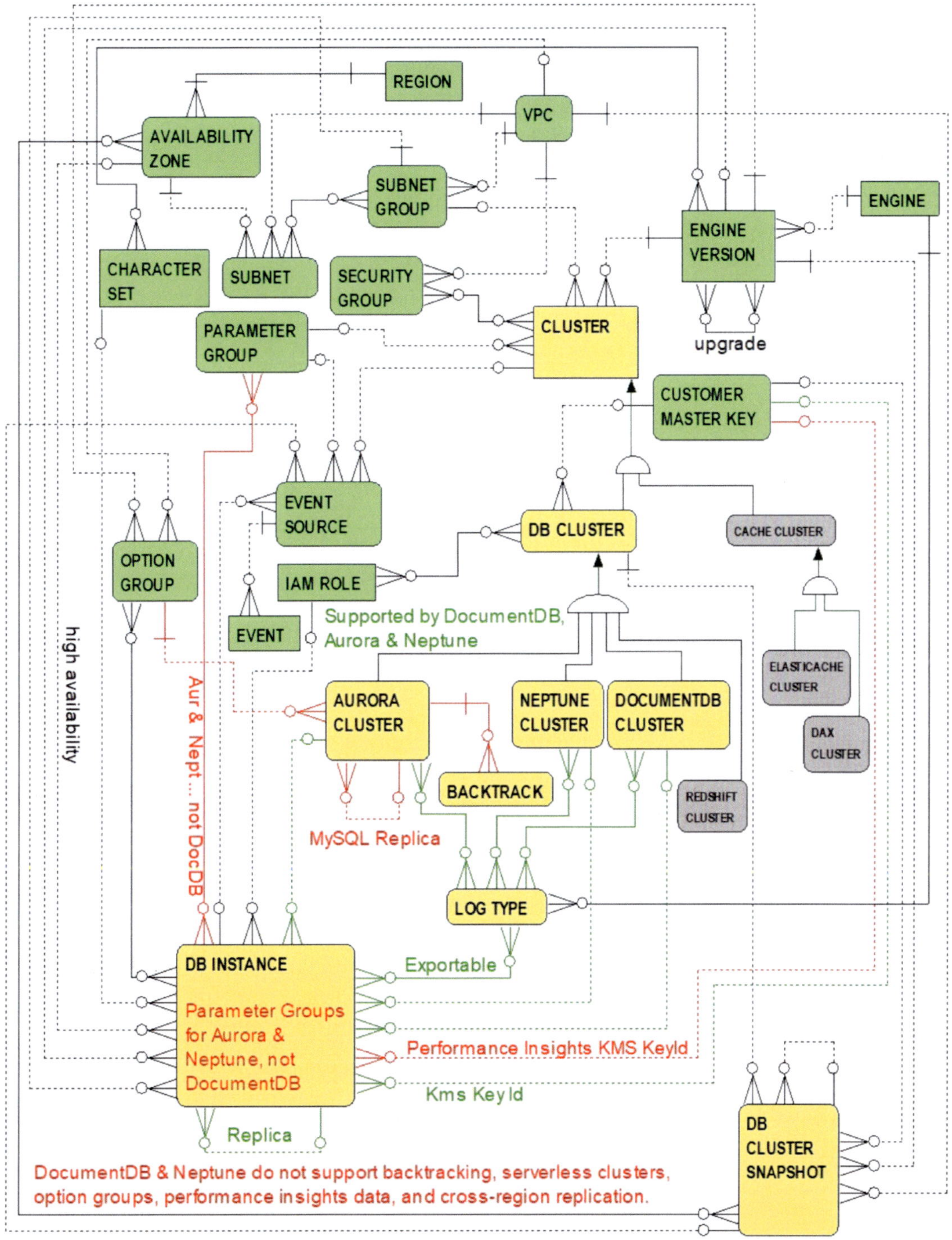

Figure 5.4 DocumentDB Summary Compared with Aurora, Neptune

Amazon DocumentDB is a platform-based data service offered by AWS. Like Aurora and Neptune, DocumentDB is equipped with solid-state virtual storage facilities optimized for improving database performance, reliability, and availability. DocumentDB enables high availability by synchronizing database updates to read replicas in the event of a system failure. Data replication across availability zones is also supported.

DocumentDB is an example of a *document database* compatible with MongoDB, a NoSQL database service. DocumentDB provides a non-relational approach to managing semi-structured data based on key-value pairs and document storage. DocumentDB stores semi-structured data as documents in its document database. Documents are self-describing -- composed of key-value pairs defined in a JSON-like format (BSON). A document database (compared to the relational approach) appeals to some programmers intuitively comfortable with using a document format for manipulating data in a program and then saving the same document format in a database.

DocumentDB API language concepts in contrast with SQL includes: ***Collection*** (corresponds to a relational table); ***Document*** (correspond to rows in a relational table); ***Field*** (A document is composed of one or more fields -- key-value pairs -- corresponding to a row columns); **Embedded Document** (nested key-values pairs enabling complex documents); ***Object Id*** (a document's unique identifier corresponding to a table's PK); ***Array*** (a list of multiple values, e.g., an array of character strings).

MongoDB CRUD operations include: ***insertOne*** (akin to SQL INSERT, inserts a document into a collection); ***insertMany*** (inserts multiple documents); ***find*** (akin to SQL SELECT, retrieves one or more documents); ***updateOne*** (akin to SQL UPDATE, updates a single document based on search criteria); ***updateMany*** (updates all documents satisfying search criteria); ***deleteOne*** (akin to SQL DELETE, removes a single document based on search criteria); ***deleteMany*** (removes all documents satisfying search criteria).

DocumentDB, like Neptune and Aurora, supports database clustering. DocumentDB clusters are similar to Aurora and Neptune clusters, but noteworthy differences justify differentiating DocumentDB clusters from Aurora and Neptune clusters. A DocumentDB cluster consists of a primary DB instance (for read and write operations) and as many as 15 read replica DB instances (updates are automatically synchronized from the primary). DocumentDB replicas offload data retrieval requests. DocumentDB automatically switches to a read replica if the primary DB instance crashes. DocumentDB maintains a single, virtual cluster storage volume (replicated across availability zones) which is accessible to all cluster instances. Also similar to Aurora and Neptune, a DocumentDB cluster can support one or more log types indicating the types of log files that can be published (by all DB instances in the cluster) to Amazon CloudWatch Logs.

A DocumentDB cluster, like an Aurora and Neptune cluster, is *a kind of* DB cluster which in turn is *a kind of* cluster. A DocumentDB cluster inherits a KMS customer master key (i.e., a CMK for data encryption), IAM roles, and cluster snapshots because it is *a kind of* DB cluster. A DocumentDB cluster indirectly inherits (because a DB cluster is *a kind of* cluster) many security groups, a subnet group, a parameter group, an engine version, and is a source for event notifications.

After a DocumentDB cluster is created, conventional MongoDB tools and utilities (e.g., the MongoDB Shell, MongoDB drivers and various programming language APIs) running on client EC2 instances are used for connecting and developing DocumentDB applications.

Unlike Aurora, both DocumentDB and Neptune do not support backtracking, serverless clusters, cross-region replication, option groups, and Performance Insights metrics data. Unlike Aurora and Neptune, DocumentDB does not support parameter groups for DB instances.

Chapter 6 – Amazon Redshift

Amazon Redshift Introduction

Amazon Redshift (Redshift) like RDS, Aurora, Neptune and DocumentDB, is yet another managed platform-based data service offered by AWS. Like RDS and Aurora, Redshift offers a relational database service. Redshift is especially targeted and recommended for robust applications requiring efficient access to very large data sets. Among such categories of applications include business intelligence (BI), data warehouse (DW), online analytic processing (OLAP), ETL (i.e., extract, transform, and load), machine learning and artificial intelligence. Of course RDS and Aurora engines can also be applied for such application categories, but Redshift offers advantages promoting a closer-fitting solution strategy specifically for BIDW applications in the cloud. Redshift is based on PostgreSQL and thereby facilitates application migration and minimizing changes to existing SQL-based applications.

Like Aurora, Neptune and DocumentDB, Redshift enables clusters of database instances. The Redshift vocabulary, however, slightly wanders from the others in that Redshift computing resources are called *nodes* rather than *DB instances*. The proper Redshift lingo, therefore, is to say that a Redshift cluster is collection of one or more nodes.

The concept of *nodes* is similar to *DB instances*, but nodes are sufficiently dissimilar (i.e., in terms of relationship and attribute differences) thereby justifying the introduction of a separate entity (i.e., the RS NODE entity), rather than attempting to force reuse of the DB INSTANCE entity, which of course was recycled in the previous two chapters for Neptune and DocumentDB clusters. Similarities and differences between RS nodes and DB instances will be emphasized in data modeling diagrams ensuing in this chapter. Another slight variation in the Redshift vernacular is that a Redshift cluster consists of *a leader node* assisted by one or more *compute nodes* -- rather than a *primary DB instance* synchronized with one or more *read replica DB instances* (i.e., the verbiage for Aurora, Neptune and DocumentDB clusters).

Redshift compute nodes -- unlike Aurora, Neptune and DocumentDB -- **do not** serve as read replicas. Instead, compute nodes are the engines for enabling Redshift's *massively parallel processing* (MPP) capability and, therefore, are crucial for optimizing query performance. In essence, compute nodes are the work horses for "dividing and conquering" complex queries across very large *partitioned data* sets. *Partitioned data* in this context connotes that the rows of a relational table, defined for a Redshift database, are allocated and distributed among multiple compute nodes, thereby enabling data to be accessed and processed in parallel.

For a Redshift relational table, data partitioning is optional but, curiously, is not specified via the Redshift console, CLI commands or API. Instead, Redshift supplements its support for PostgreSQL by furnishing a special syntax (i.e., a Redshift-specific SQL syntax) for enabling data partitioning. Specifically, when creating a Redshift database table (e.g., via a SQL client tool such as SQL Workbench/J or other SQL client tools compatible with PostgreSQL),[46] the SQL *CREATE TABLE* statement features a special *DISTSTYLE* parameter for specifying one of several options for distributing a table's rows. For example, this parameter can specify that data is to be partitioned by designating a column of the table as the *distribution key,* which is specified using yet another parameter (i.e., *DISTKEY (column_name)*) for the CREATE TABLE command. Values of this distribution key are the foundation for horizontally partitioning table rows among compute nodes.[47]

In addition, Redshift accomplishes data partitioning by dividing up each compute node's memory and storage capacity into two or more *slices*. Slicing enables additional parallelism to be carried out within each compute node beyond processing the workload in parallel across multiple compute nodes. Disseminating data to compute node slices and delegating the amount of work to be performed by each compute node slice -- for rapidly responding to parallel queries or other parallel data manipulation activity -- is the responsibility of the leader node. The leader node coordinates and receives the result sets from compute node slices -- each independently operating in parallel -- and aggregates final results of the operation.

Compute nodes can be scaled vertically (i.e., switching to node types with more capacity in terms of vCPUs, memory and storage) or scaled horizontally by provisioning additional compute nodes. The leader node, as the primary instance in a Redshift cluster:

- Supervises all external interaction and connections with client application programs.
- Listens for and receives queries from client applications.
- Creates *execution plans* for querying data residing on the leader node or distributed on tables across compute nodes.
- Generates code based on execution plans.
- Delegates and delivers code (i.e., SQL statements) to compute nodes with each focusing on its apportioned data.
- Coordinates and receives intermediate results from compute nodes.
- Aggregates final query results to requesting applications.

Compute nodes accomplish their delegated tasks and revert query results back to the leader node for collecting and combining a final response to the requestor.

In addition to flaunting MPP, Amazon Redshift achieves additional performance efficiencies and further distinguishes itself from RDS, Aurora, DocumentDB and Neptune by exploiting *columnar* data storage for database tables. Columnar storage represents a paradigm shift from traditional RDBMS storage strategies and is noted for boosting the performance of complex analytic queries. More specifically, for traditional RDBMS tables, column values for each row in a table are stored sequentially. One or more table rows, each with their contiguous column values, are contained in *storage blocks*. This approach enhances OLTP performance where transaction queries and updates often result in accessing or combining (i.e., *joining)* a small subset of tables and table rows. For each row that is accessed, many column values are commonly read or written in OLTP transactions. Having all column values for each row stored consecutively and sequentially in storage blocks (i.e., *row-wise* storage) is an advantage for OLTP.

BIDW and OLAP, however, are fundamentally different from OLTP. OLAP queries often result in accessing a relatively small number of columns at a time (e.g., several column *facts* or metric calculations maintained in a *Fact Table*) joined across possibly several tables (e.g., defined for a *Star Schema*).[48] For such cases, columnar storage is a closer fit and fundamentally different in that values for a single column for many rows are stored in each data block. Each storage block has the same data type and a common data compression scheme for decreasing space requirements and reducing I/O operations. Redshift needs to focus on retrieving only a smaller number of data blocks for the small number of columns necessary to answer the OLAP query. In essence, a columnar storage approach results in fewer I/O requests, less data loaded into memory, and more in-memory processing when executing queries.

In addition to championing MPP and columnar storage, Amazon Redshift achieves additional performance efficiencies -- and further distinguishes itself from RDS, Aurora, Neptune and DocumentDB -- by offering node type options for using solid-state storage devices. Like RDS and Aurora, Amazon Redshift also enables the purchase of reserved instances for compute nodes thereby offering considerable cost savings compared to on-demand pricing.

Redshift Clusters

Figure 6.1 introduces entities and relationships focusing on Redshift fundamental concepts.

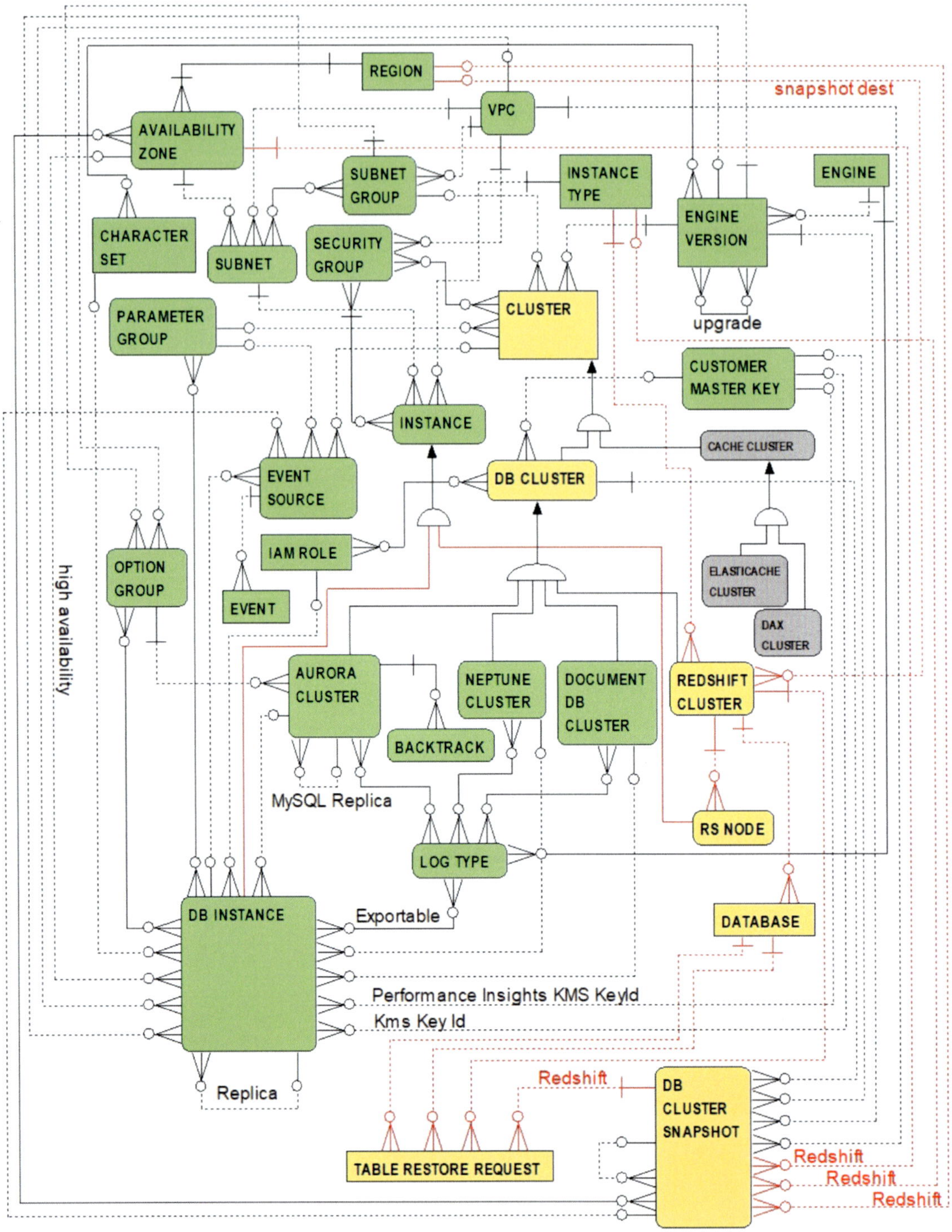

Figure 6.1 Entities and Relationships … Redshift Fundamental Concepts

Figure 6.1 highlights the REDSHIFT CLUSTER entity and reaffirms the explanation of the CLUSTER entity and its sub-type hierarchy from previous chapters. Similar to Aurora, Neptune and DocumentDB, Redshift also enables the provisioning of a cluster of instances -- referred to as *nodes* -- consisting at least of a single *leader node* and possibly many *compute nodes*. The REDSHIFT CLUSTER entity, at the same level as the AURORA CLUSTER, NEPTUNE CLUSTER and DOCUMENTDB CLUSTER entities, is a sub-type of the DB CLUSTER entity. Each Redshift cluster is *a kind of* DB cluster and, therefore, directly inherits attributes and relationships from the DB CLUSTER entity, and indirectly inherits attributes and relationships from the CLUSTER entity.

As a sub-type of the DB CLUSTER entity, a Redshift cluster (i.e., the data maintained by the leader node and compute nodes) is optionally encrypted using a KMS customer master key, and enjoys one or more IAM roles enabling Redshift to access other AWS services on behalf of the Redshift cluster. Figure 6.1 also recaps that a Redshift cluster, as *a kind of* DB cluster, can be backed up by one or more DB cluster snapshots. Unlike Aurora, Neptune and DocumentDB clusters, snapshots for a Redshift cluster can be automatically copied to a different region (i.e., *cross-region* copies) for supporting disaster recovery. Note the one-to-many relationship from the REGION entity to the REDSHIFT CLUSTER entity. This identifies the *Destination Region* (labelled "snapshot dest") to which snapshots are automatically copied if such cross-region copies are enabled for a Redshift cluster.

Note in Figure 6.1, three additional relationships (highlighted in red) peculiar to Redshift snapshots. The one-to-many relationship from the REGION entity to the DB CLUSTER SNAPSHOT entity indicates that a region can be the *source region* (i.e., the starting point for cross-region copies) for many Redshift cluster snapshots that are automatically copied to a Redshift cluster's destination region. The one-to-many relationship from the AVAILABILITY ZONE entity to the DB CLUSTER SNAPSHOT entity identifies the availability zone where the Redshift cluster was created. The one-to-many relationship from the INSTANCE TYPE entity to the DB CLUSTER SNAPSHOT entity identifies the node type for all of the Redshift cluster nodes.

Common attributes inherited by the REDSHIFT CLUSTER entity from the DB CLUSTER entity include the sample attributes noted in previous chapters and conveniently replicated here as follows:

- *backup retention period:* The number of days that automatic backups are saved prior to being purged.
- *database name:* Identifies the name of the cluster's initial database.
- *master username:* The master username for logging into the cluster's DB instances.
- *storage encrypted:* Indicates whether or not the cluster's data is encrypted.

Indirectly inherited from the CLUSTER entity, a Redshift cluster can serve as a source for event notifications, and is associated with many security groups, a single subnet group, parameter group, and engine version. This means that a cluster of Redshift nodes instances has:

- A set of security groups for controlling access to Redshift cluster instances.
- A subnet group within whose subnets Redshift cluster instances can be launched.
- A single parameter group specifying parameters for configuring all cluster instances.
- The Redshift engine version for all cluster instances.

Common attributes indirectly inherited by the REDSHIFT CLUSTER entity from the CLUSTER entity include the following sample attributes also noted in previous chapters:

- *cluster identifier:* A user-supplied cluster identifier.

- *cluster create time:* The time when the cluster was created.
- *status:* Indicates the current state of the cluster.
- *endpoint:* Specifies the cluster's access information (i.e., DNS host name and port number) enabling applications to connect to the cluster.
- *preferred maintenance window:* Identifies the time period when maintenance activities (e.g., system upgrades) are preferred.

The one-to-many relationship from REDSHIFT CLUSTER entity to the RS NODE entity indicates that each Redshift cluster is composed of one or more Redshift nodes. Like the DB INSTANCE entity, the RS NODE entity is a sub-type of the INSTANCE entity. Many of the relationships to the DB INSTANCE entity (e.g., for exportable log types, KMS keys, replicas, etc.) are not applicable to the RS nodes, thereby sufficiently justifying a separate entity for Redshift instances.

Note that Redshift documentation refers to Redshift nodes simply as "nodes" which, at first glance, would clearly suggest that the RS NODE entity in Figure 6.1 should instead simply be named the "NODE" entity. But as will be shown in subsequent chapters, Cache clusters also consist of "nodes" which are substantially different -- from a data modeling perspective -- from Redshift nodes. Hence, so as not to clash and confuse, Figure 6.1 takes the liberty of using the entity name "RS NODE" to avoid conflict with other AWS database services yet to be described (in Chapters 7 and 8).

One RS node is the leader node (i.e., primary) and the remaining RS nodes serve as compute nodes. For example, when creating a Redshift cluster via the Redshift console, a leader node is automatically created and an option is provided for specifying the number of compute nodes to be launched. The number of compute nodes can be changed using the Redshift console, CLI, or API. After a Redshift cluster is created, customary SQL client tools (i.e., tools compatible with PostgreSQL or other SQL client tools that are DBMS and platform independent) can be used to connect to the cluster database.

In addition to previously described attributes inherited from its super-type hierarchy (i.e., foreign keys and other attributes inherited from the CLUSTER and DB CLUSTER entities), and additional attributes stemming from one-to-many relationships (e.g., the cluster's instance type for the nodes in the cluster, and the destination region configured for cross-region snapshot copies), noteworthy examples of attributes peculiar to the REDSHIFT CLUSTER entity (i.e., attributes which are not evident in Aurora, Neptune or DocumentDB API documentation) include:

- *allow version upgrade*: Indicates whether or not automatic upgrades are allowed to be performed within a maintenance window for major engine versions.
- *current rate in megabytes per second:* The data transfer rate (in MB/sec) when transferring data to a new cluster during a cluster resizing activity. A Redshift cluster can be resized by modifying a cluster's node type and the number of compute nodes.
- *data transferred in megabytes:* The total amount of data that has been transferred to a new cluster during a cluster resizing activity.
- *total data in megabytes:* The total amount of data to be transferred (MB/sec) for a cluster resizing activity.

- *retention period:* For snapshots automatically cloned from a cluster in a source region, the period of time (in days) that the backup is saved in the cross region to which it was copied.

Redshift Table Restore Requests

In addition to restoring entire databases from snapshots, a feature described in Chapter 3 relative to all DB clusters, Redshift also provides the ability to restore individual database tables from database snapshots. Figure 6.2 introduces entities and relationships focusing on Redshift table restore requests.

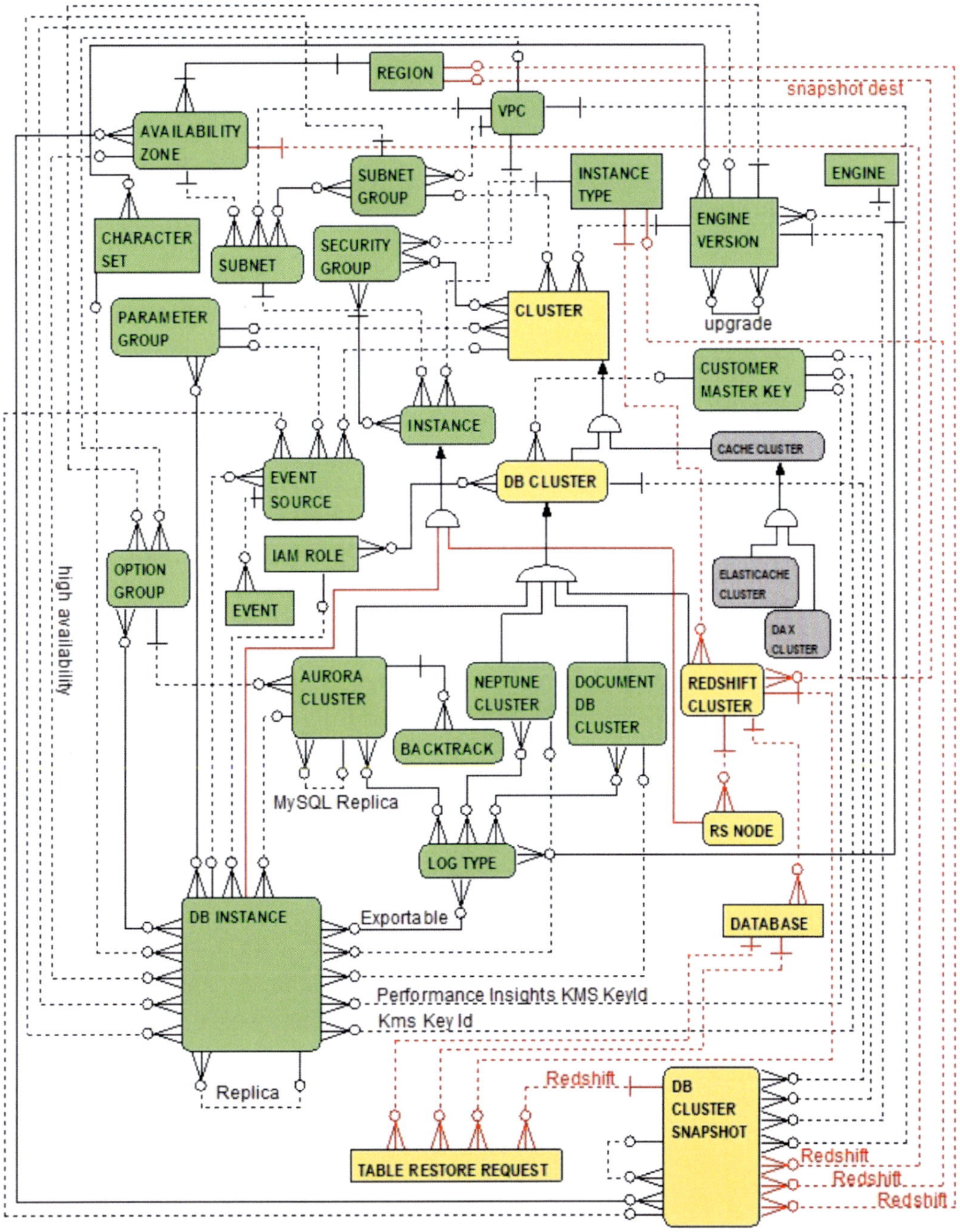

Figure 6.2 Entities and Relationships … Redshift Table Restore Requests

The one-to-many relationship from REDSHIFT CLUSTER entity to the DATABASE entity indicates that each Redshift cluster supports one or more databases. The one-to-many relationship from REDSHIFT CLUSTER entity to the TABLE RESTORE REQUEST entity shows the history of table restore requests associated with each Redshift cluster. Two one-to-many relationships from the DATABASE entity to the TABLE RESTORE REQUEST entity suggest that each request to restore a table necessarily must identify the *source database* and a *target database* involved in the recovery. The source database contains the table to be retrieved and the target database identifies the database where the table is to be restored.

The one-to-many relationship from the DB CLUSTER SNAPSHOT entity to the TABLE RESTORE REQUEST entity identifies the snapshot from which the source table will be snatched. Note in Figure 6.2 that this relationship is specially annotated emphasizing peculiarity to Redshift snapshots (i.e., not applicable to Aurora, Neptune and DOCUMENTDB cluster snapshots) and, therefore, was excluded from diagrams illustrating the DB CLUSTER SNAPSHOT entity in previous chapters. Industrious data modeling students may wish to ponder the possibility of modifying Figure 6.2 by introducing a new sub-type of the DB CLUSTER SNAPSHOT entity depicting all relationships peculiar to Redshift snapshots. Recall from the explanation of a conceptual data model (CDM) and a more detailed logical data model (LDM) in the Introduction, that such an exercise might arguably be more appropriate and justifiable when evolving to an LDM from a CDM.

Noteworthy attributes of the TABLE RESTORE REQUEST entity include the *source table name (*i.e., the table being recovered from source database in the snapshot) and *new table name* (i.e., the name of the new table to be constructed in the target database resulting from the recovery request).

Redshift, DocumentDB, Neptune and Aurora Cluster Comparison

Clearly from the previous sections, Redshift clusters are similar to Aurora, Neptune and DocumentDB clusters. As noted, each of these clusters is *a kind of* DB cluster (i.e., the REDSHIFT CLUSTER, DOCUMENTDB CLUSTER, AURORA CLUSTER and NEPTUNE CLUSTER entities are sub-types of the DB CLUSTER entity) and as such enjoy all of the relationships and attributes directly inherited from their parent DB CLUSTER entity and indirectly inherited from their grandparent CLUSTER entity. A Redshift cluster, however, exhibits several noteworthy differences which justify establishing the REDSHIFT CLUSTER entity as a distinct sub-type of the DB CLUSTER entity. Specifically …

- DocumentDB, Aurora and Neptune clusters each have a one-to-many relationship to the DB INSTANCE entity and share similar concepts and vocabulary. Each of these clusters is composed of one or more DB instances, where one DB instance is primary and the remaining servers are read replicas. Instead, Redshift clusters are composed of nodes -- a leader node and possibly many compute nodes. The relationship to the DB INSTANCE entity was not recycled for Redshift clusters.
- DocumentDB, Aurora and Neptune clusters each have a many-to-many relationship to the LOG TYPE entity indicating their support for publishing log

files to Amazon CloudWatch Logs. Redshift does not support exportable log types.

- Unlike Aurora, Redshift -- in addition to DocumentDB and Neptune -- does not support backtracking, serverless clusters, cross-region replication, and option groups.
- Redshift clusters -- unlike DocumentDB, Neptune, and Aurora -- support the ability to individually restore database tables from a cluster snapshot without requiring a full restore of the cluster.
- The RS NODE entity is substantially different from the DB INSTANCE entity. The one-to-many reflexive relationship from the DB INSTANCE entity to itself, indicating that many DB instances that can serve as read replicas for a primary DB instance, is not applicable to RS nodes. Redshift compute nodes are not read replicas for the leader node. As previously noted, several other relationships to the DB INSTANCE entity (e.g., for exportable log types, KMS keys, high availability, event sources, etc.) are not applicable to the RS nodes) copiously justifying a separate RS NODE entity for Redshift instances.

Chapter Summary

Figure 6.3 extends Figure 6.2 with additional commentary summarizing Amazon Redshift basic concepts along with Aurora, Neptune, and DocumentDB similarities and differences.

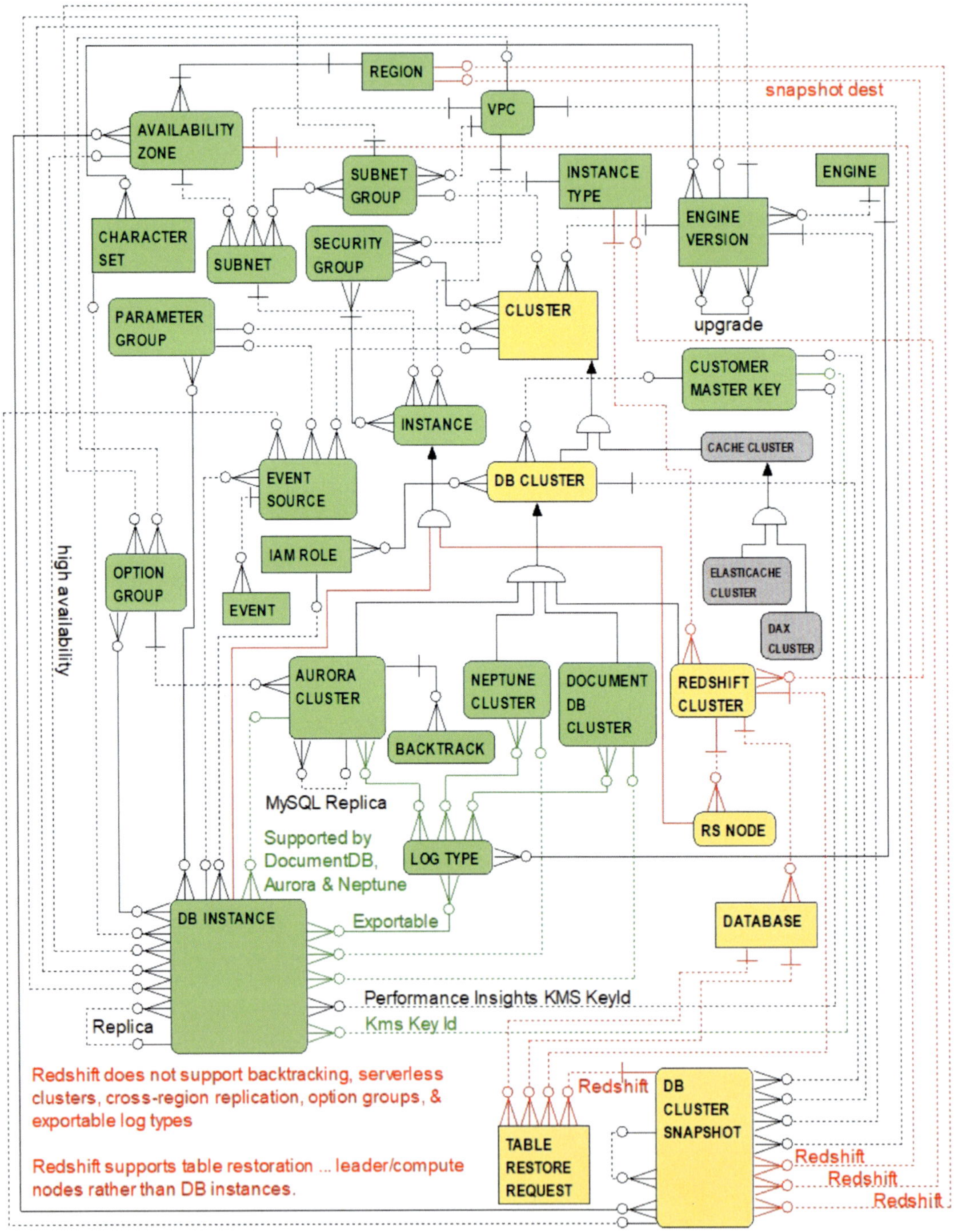

Figure 6.3 Redshift Summary Contrasting Aurora, Neptune and DocumentDB

Amazon Redshift (Redshift) -- like RDS, Aurora, Neptune and DocumentDB -- is another managed, platform-based database service offered by AWS. Like RDS and Aurora, Redshift offers a relational database service, based on PostgreSQL, and appeals to apps with very large data sets (e.g., BIDW, OLAP, ETL, and AI). Like Aurora, Neptune and DocumentDB, Redshift enables clusters of database instances. Redshift clusters are composed of *nodes* rather than *DB instances*. A Redshift cluster consists of *a leader node* assisted by one or more *compute nodes*, rather than using the terminology for other DB clusters … a *primary DB instance* with one or more *read replica DB instances*.

Redshift compute nodes -- unlike Aurora, Neptune and DocumentDB -- **do not** serve as read replicas. Compute nodes enable *massively parallel processing* (MPP) crucial for optimizing query performance across very large *partitioned data* sets (i.e., table rows are distributed across compute nodes). A Redshift-specific SQL syntax enables data partitioning. Redshift's *CREATE TABLE* statement features a *DISTSTYLE* parameter for specifying options for distributing a table's rows (e.g., by designating a table column as the *distribution key* for horizontal partitioning).

Redshift accomplishes data partitioning by dividing each compute node's memory and storage capacity into two or more *slices*. This enables additional parallelism carried out within each compute node. The leader node disseminates data to compute node slices and delegates work (i.e., code based on execution plans) to be performed by each compute node slice. The leader node coordinates and aggregates result sets from compute node slices.

Redshift achieves additional OLAP performance efficiencies by exploiting *columnar* data storage for database tables. OLAP queries often access a small number of columns at a time (e.g., in Dimension and Fact Tables for Star Schemas). Values for a single column for many rows are stored in each data block, rather than row-wise data blocks benefitting OLTP. Columnar data blocks take advantage of a common data type and compression scheme thereby decreasing space requirements and reducing I/O operations.

Redshift also offers node type options for using solid-state storage devices and -- like RDS and Aurora -- enables the purchase of reserved instances for compute nodes offering considerable cost savings compared to on-demand pricing.

Redshift clusters are similar to Aurora, Neptune and DocumentDB clusters -- but noteworthy relationship and attribute differences justify differentiating Redshift clusters from the other DB clusters. A Redshift cluster -- like an Aurora, Neptune and DocumentDB cluster, is *a kind of* DB cluster which in turn is *a kind of* cluster. A Redshift cluster inherits a KMS customer master key (i.e., a CMK for data encryption), IAM roles, and cluster snapshots because it is *a kind of* DB cluster. A Redshift cluster indirectly inherits (because a DB cluster is *a kind of* cluster) many security groups, a subnet group, a parameter group, an engine version, and is a source for event notifications. After a Redshift cluster is created, customary SQL client tools (i.e., tools compatible with PostgreSQL or other SQL client tools that are DBMS and platform independent) can be used to connect to the cluster database.

Unlike other DB clusters …

- Snapshots for a Redshift cluster can be automatically copied to a different region (i.e., *cross-region* copies) for supporting disaster recovery.

- Redshift provides the ability to individually restore database tables from database snapshots and maintains the history of table restore requests.
- Redshift does not support exportable log types.
- Redshift -- unlike Aurora, but in addition to DocumentDB and Neptune -- does not support backtracking, serverless clusters, cross-region replication, and option groups.

Chapter 7 – Amazon DynamoDB

Amazon DynamoDB Introduction

Amazon DynamoDB (DynamoDB) -- like RDS, Aurora, Neptune, DocumentDB and Redshift -- is yet another managed, platform-based database service offered by AWS. Like Neptune and DocumentDB, DynamoDB is a NoSQL database engine providing a non-relational approach to managing data. DynamoDB supports both structured and semi-structured data based on key-value pairs and document storage. This includes native support for JSON documents and the ability to access and manipulate data internal to JSON documents.

As fully expected from a managed AWS service, DynamoDB minimizes sleep deprivation from worry surrounding operational issues and run-of-the-mill admin tasks, such as provisioning hardware, performing backups, restoring databases, installing, upgrading and patching software, etc. Like Neptune, DynamoDB offers improved performance, availability and reliability by using solid state storage devices (SSDs), partitioning data among SSDs, and balancing workload across server resources as necessary to satisfy latency and storage needs. Also like Neptune, DynamoDB transparently replicates data across multiple availability zones and automatically expands and repartitions storage as necessary without service interruption.

Substantially distinguishing itself from previously described AWS database services, DynamoDB exceeds typical platform-based service expectations by virtue of being totally serverless. This means that DynamoDB shields customers from many details regarding the provisioning of appropriate virtual server resources for adequately anticipating database workloads. DynamoDB database instances are not explicitly provisioned as for RDS, Aurora, Neptune, DocumentDB and Redshift. Primary DB instances, replica DB instances, leader nodes and compute nodes are irrelevant to DynamoDB and need not be contemplated as for the previously described database services. The need for upfront capacity planning, especially for deploying new applications -- where transaction rates are unpredictable, fluctuating or cyclic -- is minimized.

DynamoDB is ideally targeted and commonly considered for implementing globally distributed, *internet scale* applications. Such applications (e.g., e-commerce shopping carts, global gaming applications, on-line bidding, etc.) can go viral and may require the ability to scale compute and storage resources for potentially supporting tens of thousands of concurrent international customers -- collectively generating tens of thousands of database requests per second and amassing enormous data sets.

In such cases, DynamoDB optionally simplifies capacity management by automatically increasing or decreasing compute and storage resources to accommodate on-demand database traffic and dependably maintain responsive performance. The paradigm shift for DynamoDB is that the customer specifies capacity in terms of *throughput requirements* (i.e., the anticipated number of reads and writes) instead of specifying the number of database instances needed to be scaled up or down.

Although categorized as a NoSQL approach, DynamoDB does support the concept of *tables* analogous to relational tables (more detail about DynamoDB tables later). Throughput requirements are specified in terms of the anticipated number of reads and writes on DynamoDB tables. A table's read/write throughput capacity can be increased or decreased. DynamoDB responds by automatically reserving (transparent to the customer) sufficient servers and appropriately spreading database requests, such that a table's throughput and response time requirements are satisfied. Data partitions are also automatically and appropriately allocated -- another background activity -- to support additional storage and throughput requirements.

Simply stated, for basic DynamoDB capacity planning, throughput requirements for each table are specified when a table is created (i.e., *provisioned throughput* -- the expected I/O traffic in terms of the number of reads and writes per second). Separate throughput settings for expected read activity (i.e., *read capacity units*, RCUs) and write activity (i.e., *write capacity units*, WCUs) are specified for each table. These capacity settings are treated by DynamoDB as the basis for allocating appropriate resources for sustaining the projected I/O activity. RCUs and WCUs are considered to be maximum values. If I/O traffic against a table exceeds the specified RCUs or WCUs, DynamoDB can regulate traffic by *throttling* and retrying database requests. Such is transparently performed by DynamoDB without the need for implementing special application logic.

For auto scaling, DynamoDB exploits scaling policies (i.e., *auto scaling policies*) for automatically adjusting provisioned throughput settings based on actual utilization (i.e., a table's actual read/write traffic). For this option, upper and lower limits are specified for a table's RCUs and WCUs. In addition, a *target utilization percentage* is specified. Auto scaling basically means that actual table traffic is monitored and compared with the utilization target. I/O table traffic can increase gradually or swiftly requiring an automatic increase of read or write capacity to ensure response time satisfaction.

Contrarily, I/O table traffic can decrease gradually or abruptly requiring an automatic decrease of read or write capacity, thereby avoiding the cost of idle capacity. Auto scaling appropriately scales throughput settings up or down consistent with the desired target utilization (more detail about auto scaling later). Auto scaling is very appropriate for new apps with ambiguous traffic requirements and apps having reasonably well-known traffic patterns with gradual or predictable shifts up or down in concurrent usage (e.g., based on time zones).

DynamoDB -- like RDS, Aurora, and Redshift -- also provides cost saving opportunities by allowing for reserved capacity. However -- unlike RDS, Aurora, and Redshift where the purchase of reserved instances is offered -- DynamoDB offers the upfront purchase of RCUs and WCUs. The benefit of course is cost savings derived from a one-time upfront price compared to on-demand pricing. The tradeoff is similar to RDS, Aurora, and Redshift in that a long term usage commitment is required.

Also implied by use of the term NoSQL, DynamoDB's non-relational approach connotes that it supports a data manipulation language other than SQL for creating, reading, updating and deleting data (i.e., fundamental CRUD data operations) maintained within its databases. The terminology and basic concepts for DynamoDB and relational databases are comparable. Both approaches involve the definition of database *tables*. For RDBMS approaches, tables are ordinarily created in terms of predefined *columns* whose values are maintained in one or more table *rows*. Instead of rows and columns,

DynamoDB tables are composed of one or more *items* maintaining values of *attributes*. Each attribute has a name and a value. Attributes and values are maintained within each item as a collection of name-value pairs. Special document data types (e.g., the "map" data type) are also supported for nesting collections of attributes (i.e., for supporting JSON files).

Vocabulary differences aside, however, a more striking distinction is the idea that DynamoDB tables are *schemaless* and, thereby, provide a greater degree of flexibility for defining tables and inserting items into a table. For establishing uniqueness, a primary key, consisting of either one or two attributes, is required when defining DynamoDB tables. Other than the primary key, attributes and their data types are not required to be defined up front at the time a DynamoDB table is being created. Unlike rows in relational tables, each item in a DynamoDB table can flexibly consist of its own, private set of different attributes and data types. Attributes can be dynamically added to an item when inserting items into a table. In an RDBMS, names of columns along with their data types and sizes are usually defined when creating a relational table. After its creation a relational table can be altered by adding new columns or changing existing columns, but the columns apply to all rows in a relational table.

CRUD data operations for DynamoDB vs. SQL are summarized as follows:

- **Create**:
  - SQL: An *INSERT* statement adds a row to a table
  - DynamoDB: The *PutItem* action adds an item to a table
- **Read**:
  - SQL: A *SELECT* statement accesses and retrieves rows from one or more tables.
  - DynamoDB: The *GetItem* action is used for retrieving a single item by its primary key. A *Query* action enables retrieval of multiple items based on query filters.
- **Update**:
  - SQL: An *UPDATE* statement is used for modifying one or more rows
  - DynamoDB: An *UpdateItem* action is used to update a single item
- **Delete**:
  - SQL: A *DELETE* statement deletes one or more rows from a table.
  - DynamoDB: The *DeleteItem* operation deletes one item at a time.

DynamoDB Basic Concepts

Figure 7.1 introduces entities and relationships focusing on DynamoDB fundamental concepts.

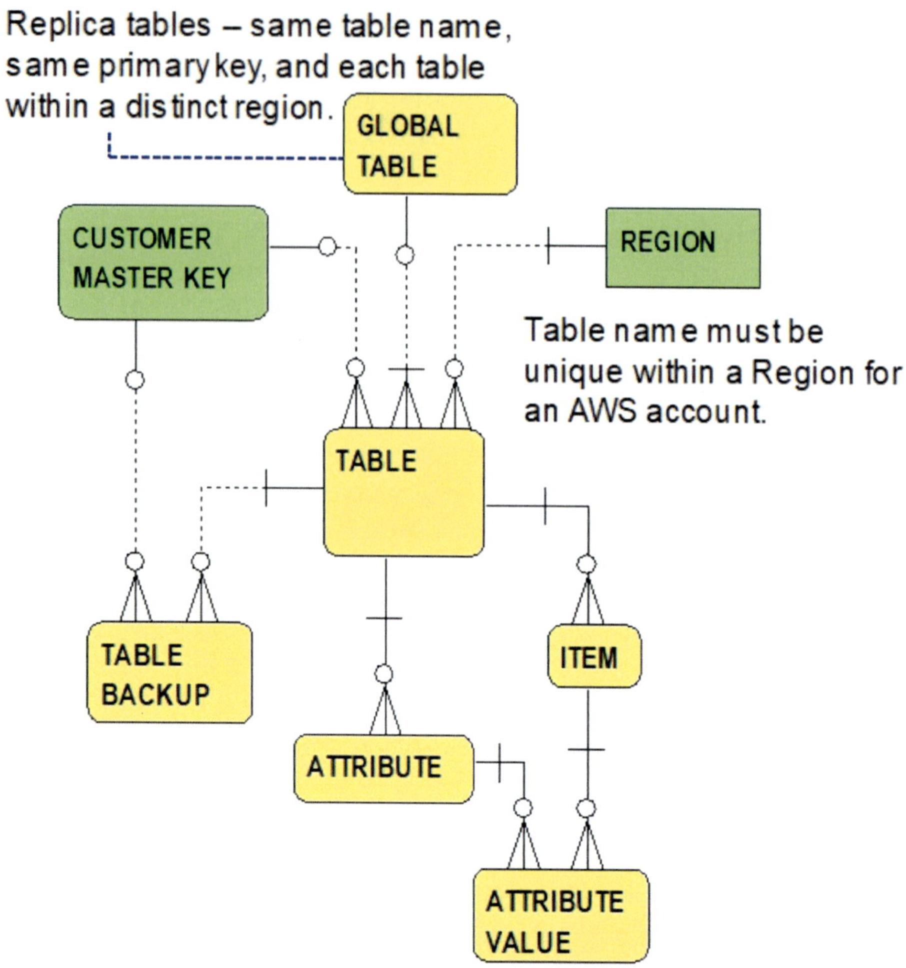

Figure 7.1 DynamoDB Fundamental Concepts

From Figure 7.1, the TABLE entity represents the set of all DynamoDB tables created via the DynamoDB console, API or CLI commands. The one-to-many relationships from the CUSTOMER MASTER KEY and REGION entities to the TABLE entity respectively depict that each DynamoDB table can optionally be encrypted and is associated with a specific region. The one-to-many relationships from the TABLE entity to the ATTRIBUTE and ITEM entities respectively indicate that each table has one or more attributes and one or more items. As many as two attributes can be designated as participating in the primary key for a DynamoDB table.

- If only one attribute is designated as the primary key, the attribute is referred to as a *partition key* (i.e., a simple primary key). A table's partition key is the basis for DynamoDB's partition management and data distribution across servers as mentioned in the previous section. The value of a partition key attribute serves as input to the DynamoDB hashing algorithm for determining the actual partition in which the item will be stored or from which it can be accessed for future retrieval or modification.
- If two attributes are designated as the primary key, they are respectively referred to as a *partition key* and *sort key* (i.e., a composite primary key).

Resorting back to conceptual data modeling parlance (recall from Chapter 1, the description of an *attribute* of an *entity* in a data model) and attempting to minimize terminology confusion (between data modeling concepts and the DynamoDB

ATTRIBUTE entity), let it be emphasized that data modeling *entities* have *attributes*. Therefore the ATTRIBUTE entity in Figure 7.1 also has entity attributes (i.e., attributes of the ATTRIBUTE entity), which in data management circles is an example of the concept of *metadata* (i.e., data about data). Therefore, from a conceptual data modeling perspective, a *key indicator* attribute of the ATTRIBUTE entity could serve as a flag indicating whether an attribute of a DynamoDB table is either a *partition key* or a *sort key* or *neither*.

The ATTRIBUTE VALUE entity resolves a many-to-many relationship between ITEM and ATTRIBUTE entities and emphasizes the *schemaless* nature of DynamoDB tables in that each item flexibly maintains data values for its own distinct set of one or more table attributes. An item of a table need not have values for all table attributes.

The one-to-many relationships from the TABLE entity to the TABLE BACKUP entity reveals that each DynamoDB table can optionally have one or more backups.

Global Tables

From Figure 7.1, the GLOBAL TABLE entity introduces DynamoDB support for *global tables,* a mechanism for automatically replicating and synchronizing information across multiple regions. The term "global table" is somewhat ambiguous in that a global table is not an actual, singular, distinct DynamoDB table. A global table is actually made up of a <u>group of replica tables</u> each having the same table name, the same primary key, and belonging to the same AWS account -- but each replica table is located in a different region.

Implementing global tables is indeed a noteworthy feature for replicating a database consisting of numerous tables across regions and, in effect, creating a *multimaster* database. Information created or updated in any one of these replica tables is automatically synchronized and propagated to all other tables in the global table's group of tables. The global table group of tables is managed by DynamoDB and dealt with as a single, composite unit. The concept of global tables is indeed a remarkable feature for obviating do-it-yourself application development effort for implementing a comparably seamless capability across regions. Other noteworthy benefits, of course, are opportunities for improved high availability, response time performance, world-wide customer access, disaster recovery, and business continuity.

In Figure 7.1, the one-to-many relationship from the GLOBAL TABLE entity to the TABLE entity indicates that each global table is simply a cluster of tables. The relationship is annotated to recall the constraints that the grouped tables are like-named, identified by the same primary key, and that each table resides in its own distinct region. A global table is created (e.g., via the DynamoDB console) simply by specifying the name of an existing table and the AWS regions to which the table is replicated and between which data updates are communally propagated.

In addition to identifiers for customer master key, region, and global table (i.e., foreign keys), noteworthy additional attributes of the TABLE entity include:

- *table name*: Maintains the table name.
- *table size bytes*: The periodically updated total table size in bytes.
- *table status*: Maintains the table's current status, e.g. "ACTIVE", "CREATING", etc.
- *read capacity units* (RCUs): Specifies the capacity requirement (aka *provisioned throughput*) for a table's ability to support a category of read operations known as *strongly consistent reads*. This is expressed as an upper limit of such read requests permitted by DynamoDB per second. Strongly consistent reads ensure that all the latest data updates are included in the query result. If this maximum is exceeded, a *throttling exception* is returned to the application by DynamoDB.
- *write capacity units* (WCUs): Similar to read capacity units, this specifies the capacity requirement for a table's ability to support write operations. This is also expressed as an upper limit of write requests permitted per second (i.e., for writing less than 1 KB for a table item). Exceeding this upper limit also results in a throttling exception.
- *last increase datetime*: Read or write capacity can be increased during the day to accommodate increased workload on a table. This attribute reflects the date and time at which the read or write throughput capacity for this table was last increased. For example, throughput capacity can be dynamically adjusted via the DynamoDB *Auto Scaling* feature which is more fully described in a subsequent section.
- *last decrease datetime*: Read or write capacity can be decreased (e.g., via DynamoDB Auto Scaling) to accommodate a decrease of workload on a table. This attribute reflects the date and time at which the read or write throughput capacity for this table was last decreased.
- *enabled: A* true/false indicator denoting whether data encryption is in effect for the table.

Global and Local Indexes

In addition to providing for a simple or composite primary key for each table, DynamoDB also allows indexes enabling alternate access paths for accelerating query latency. Figure 7.2 expands Figure 7.1 by focusing on entities and relationships for establishing DynamoDB table indexes.

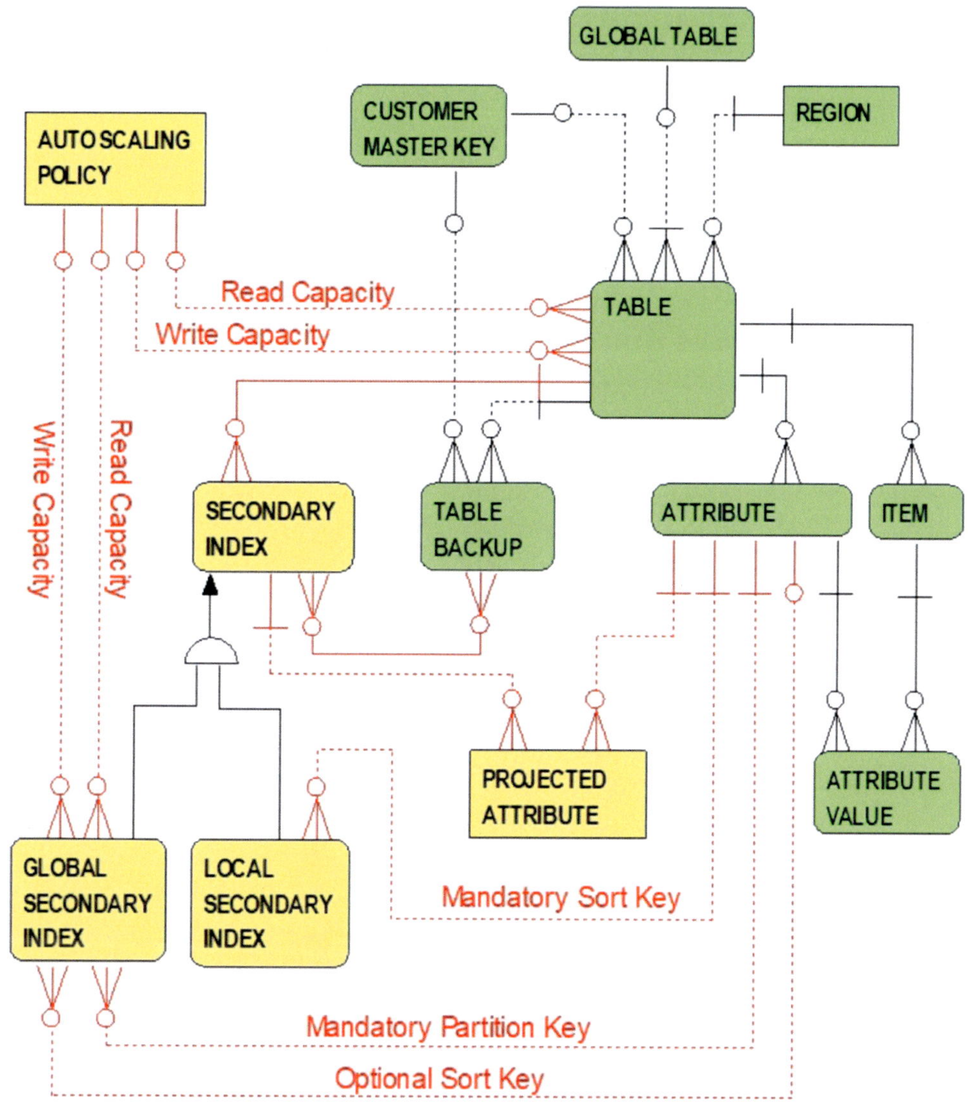

Figure 7.2 Entities and Relationships … DynamoDB Index Concepts

From Figure 7.2, the one-to-many relationship from the TABLE entity to the SECONDARY INDEX entity depicts that each DynamoDB table can be defined as having one or more secondary indexes for directly accessing table items. Note also that the SECONDARY INDEX entity is a super-type that has two sub-types represented by the GLOBAL SECONDARY INDEX and LOCAL SECONDARY INDEX entities. A secondary index is either a *global secondary index* or a *local secondary index*. Similar to a table having a primary key (i.e., simple or composite), each secondary index is also defined having up to two key attributes, referred to as a *partition key* and a *sort key*.

The rules for defining partition keys and sort keys differ for global indexes and local indexes, hence justifying their respective sub-types. The one-to-many relationships -- two relationships -- from the ATTRIBUTE entity to the GLOBAL SECONDARY INDEX entity are annotated indicating that each global secondary index has a mandatory partition key and an optional sort key (i.e., up to two attributes participating in its primary key). These can be any two table attributes.

The one-to-many relationship from the ATTRIBUTE entity to the LOCAL SECONDARY INDEX entity indicates that each local secondary index has a mandatory sort key. The partition key for each local secondary index is the same partition key as specified for the table. To emphasize, all local secondary indexes defined for a table have the same partition key as the table itself. Note well, therefore, that in a conceptual data model, no explicit relationship (i.e., from the ATTRIBUTE table to the LOCAL SECONDARY INDEX entity) is required to show this because the table's partition key can be derived from the table's attributes. As noted in a previous section, it is assumed that the table attribute serving as the partition key is "flagged" as such (i.e., there is a boolean attribute of the ATTRIBUTE entity for flagging a partition key). Please appreciate the attempt to accurately explain usage of the term "attribute" in this seemingly circular and perhaps confusing context.

Note also that the term "global" secondary index emphasizes that queries use a global secondary index when requiring access to all data partitions set up for the table. In essence, a global secondary index enables global access over all table partitions. The term "local" secondary index implies that a query using a local secondary index is restricted to accessing a single partition, which is specified by the value of the partition key in the query.

The PROJECTED ATTRIBUTE entity resolves a many-to-many relationship between SECONDARY INDEX and ATTRIBUTE entities which clarifies that additional attributes, along with the partition key attribute and sort key attribute, can be stored within (i.e., *projected* onto) an index. This enables data retrieval from the index itself as a short cut for satisfying some queries and, therefore, obviating access to the base table. At the time of defining a global or local secondary index, the additional attributes to be cloned from the base table into the index (i.e., along with partition and sort keys) can be specified. Each secondary index can include many projected attributes and, in the other direction, an attribute can appear as a projected attribute within many secondary indexes.

Auto Scaling

Again from Figure 7.2, the AUTO SCALING POLICY entity showcases the DynamoDB feature enabling automatic scaling (i.e., increasing or decreasing) of compute resources based on provisioned throughput settings (i.e., read capacity units and write capacity units). DynamoDB can focus this capability on each table separately. The one-to-many relationships from the AUTO SCALING POLICY entity to the TABLE entity portrays that two *auto scaling policies* can be associated with each table for auto scaling purposes. Each is annotated to indicate that one policy can be established for scaling read capacity and another for scaling write capacity. For example, when a Dynamo customer wishes to enable auto scaling for an existing table, the DynamoDB console provides an "Auto Scaling" option. This enables the user to specify various auto scaling parameters, such as *target utilization* and lower and upper boundaries for a table's read or write capacity. As noted previously, the target utilization is the percentage of desired capacity utilization upon which auto scaling activity is based. For example …

- If target utilization for a table's read activity is specified as 75% and …
- If the actual read traffic exceeds 75% of a table's maximum read capacity units (i.e., the read workload exceeds the target for several minutes) then …
- The table's read capacity units value is appropriately increased to approximate the desired 75% level of utilization.

This is referred to as a *scale out* activity. Similarly -- in the opposite direction -- if the actual the read workload sinks below the target utilization for several minutes, then the table's read capacity is appropriately reduced to approximate the desired 75% level of target utilization. This is referred to as a *scale in* activity. Essentially, RCU and WCU settings can be dynamically boosted or depressed based on actual utilization.

An auto scaling policy is actually a concept germane to the AWS Application Auto Scaling service, yet another AWS service.[49] The DynamoDB console, CLI, API and SDKs appropriately interface with AWS Application Auto Scaling for establishing auto scaling policies and registering tables as *target resources* for auto scaling.

Metrics pertaining to the table's actual reads and writes (i.e., *consumed read capacity units* and *consumed write capacity units*) are exported by DynamoDB to AWS CloudWatch. AWS CloudWatch initiates an alarm and calls on Application Auto Scaling after it detects that the target utilization has been breached. The auto scaling policy is examined and Application Auto Scaling then takes action by issuing the appropriate API call to DynamoDB (i.e., an Update Table API action) to alter read or write capacity.

Sample noteworthy attributes of the AUTO SCALING POLICY entity include:

- *policy name:* Specifies the scaling policy's name.
- *target value*: The target percentage of application workload to capacity that DynamoDB attempts to maintain by automatically triggering scale in or scale out actions.
- *scale out cool down*: To discourage excessive and unnecessarily frequent scale out, this specifies the wait time after a scale out event before the next scale out can occur (i.e., the minimum time between scale out events). After a scale out activity finishes, the next possible scale out cannot be initiated prior to completing the cool down time length.
- *scale in cool down*: The opposite of *scale out cool down.* To discourage excessive and unnecessarily frequent scale in, this specifies the wait time after a scale in event before the next scale in can occur (i.e., the minimum time between scale in events). After a scale in activity finishes, the next possible scale in cannot be initiated prior to completing the cool down time length.

Auto scaling policies also apply to global secondary indexes as well as tables. Mimicking auto scaling behavior for a table, auto scaling for a global secondary index basically means that the actual workload on a global secondary index (i.e., actual reads and writes to a global secondary index) can be similarly monitored and compared with a specified utilization target. Provisioned throughput capacity is also specified for a global secondary index. These settings are independent and separate from capacity settings for the associated table. Auto scaling in this case reacts to gradual or abrupt utilization changes for the index by tweaking its provisioned throughput capacity settings to guarantee that the actual load is close to the preferred target utilization.

Note also from Figure 7.2, the one-to-many relationships from the AUTO SCALING POLICY entity to the GLOBAL SECONDARY INDEX entity portray that two auto scaling policies can be associated with each global secondary index for auto scaling purposes. Each is annotated to indicate that one policy is for scaling read capacity and another for scaling write capacity. For these auto scaling policies, each global secondary index must be registered with AWS Application Auto Scaling as a target resource for auto scaling.

DynamoDB Streams

DynamoDB provides an optional table feature known as DynamoDB *Streams* which are implemented for what is generally known in the IT industry as change data capture (CDC).[50] Simply stated, DynamoDB streams are established for tracking and capturing any changes (i.e., inserts, updates, deletions) to table items. The captured changes may then be used for triggering subsequent actions. Plausible use cases include synchronizing data changes among replica tables (e.g., as previously alluded to for global tables), automatically sending email notifications alerting that certain events have occurred, maintaining logs for reviewing change history, etc.

DynamoDB captures data changes by sequentially storing a *stream record* for every modification to a table item. Within each stream record, in addition to capturing a table item's primary key, a stream can also record "before" as well as "after" modifications to table items. Stream records are clustered into groups known as *shards* which are automatically created and deleted as necessary by DynamoDB. Each stream record has a *sequence number* attribute. In order to access one or more stream records, it is necessary to know the shard to which they belong. For example, each shard has attributes indicating the range of sequence numbers (i.e., *starting sequence number* and *ending sequence number)* for facilitating access to stream records and iteration through any subset of the stream records governed within the shard.

Figure 7.3 expands Figure 7.2 by focusing on entities and relationships for DynamoDB streams.

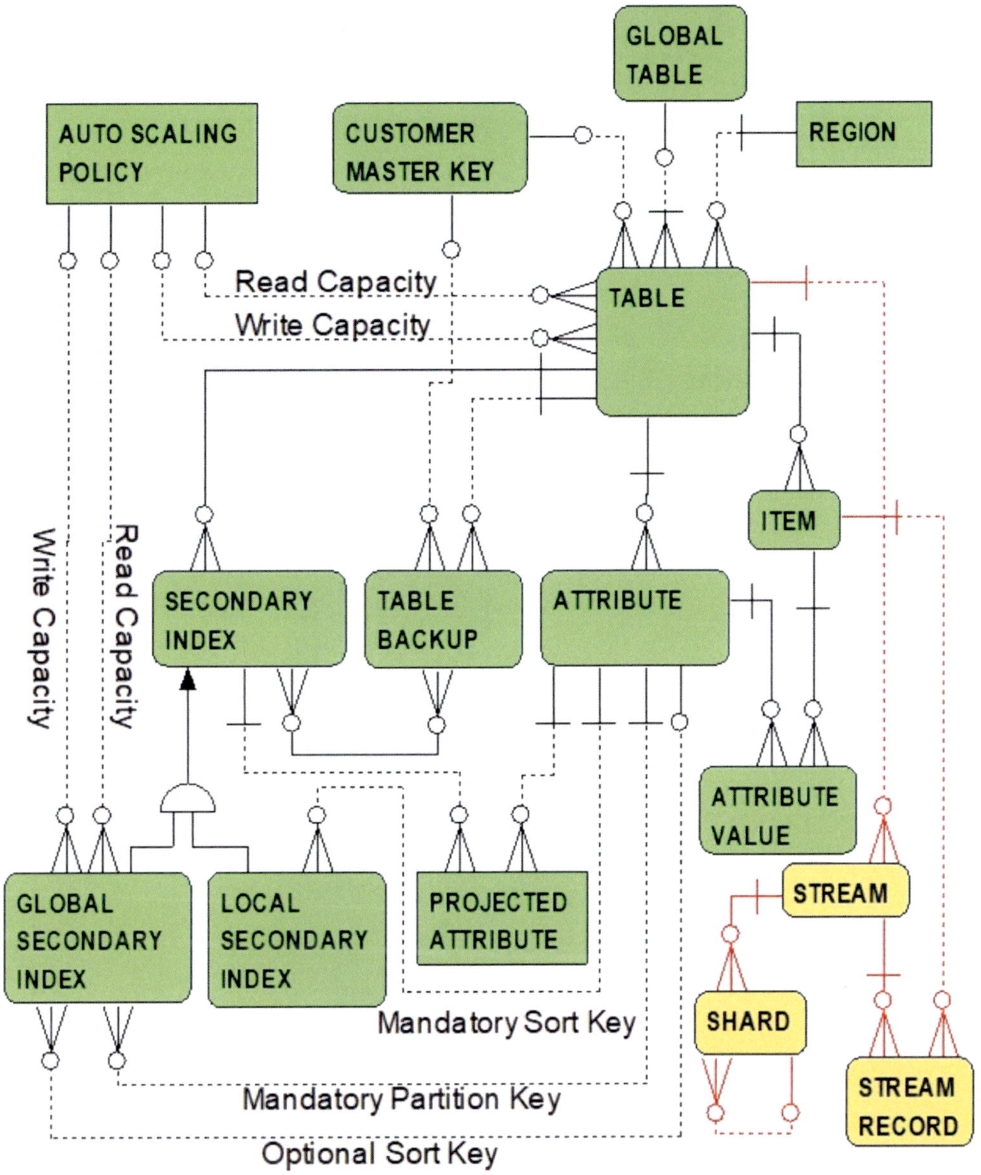

Figure 7.3 Entities and Relationships … DynamoDB Streams

From Figure 7.3, the STREAM entity showcases the DynamoDB stream capability. The one-to-many relationship from the TABLE entity to the STREAM entity indicates that each table can potentially have several streams created for capturing data changes. Only one stream, however, can be active at any point in time for a table. The one-to-many relationship from the STREAM entity to the STREAM RECORD entity portrays that each stream has many stream records each of which logs an actual change event for a specific item. The one-to-many relationship from the ITEM entity to the STREAM RECORD entity indicates that each table item potentially has many change events that have been logged for it. Each stream record must be associated with a single item.

The one-to-many relationship from the STREAM entity to the SHARD entity reminds that each stream brackets the changes into shards. Note the one-to-many reflexive relationship from the SHARD entity to itself. This suggests a hierarchical parent/child relationship which reveals that each shard (i.e., a parent shard) can possibly be automatically separated into one or more new child shards. Splitting of shards can occur as a result of a high volume of table changes. This also enables parallel processing by allowing simultaneous access to multiple shards. Note that the association of specific stream records clustered within a shard is inferred by comparing the stream record's *sequence number* to a shard's *starting sequence number* and *ending sequence number* (i.e., rather than adding a direct one-to-many relationship between the SHARD and STREAM RECORD entity).

An attribute of the STREAM entity, *stream view type,* indicates the scope of item data copied into stream records. Different valid values of the *stream view type* attribute specify whether …

- Only the key attributes of the updated item are stored in the stream record (i.e., *stream view type* = KEYS_ONLY)
- Only the *after image* of the updated item, as it appears after being updated, is stored in the stream record (i.e., *stream view type* = NEW_IMAGE)
- Only the *before image* of the updated item, as it appeared before being updated, is stored in the stream record (i.e., *stream view type* = OLD_IMAGE)
- Both the *before image* and the *after image* of the updated item are stored in the stream record (i.e., *stream view type* = NEW_AND_OLD_IMAGES)

DynamoDB Accelerator (DAX)

DynamoDB Accelerator (DAX) is a *caching* service providing an AWS option for deploying applications that are capable of delivering remarkably rapid response times and possible additional cost savings for certain use cases. For some use cases, the response times of read workloads can be reduced by an order of magnitude possibly achieving sub millisecond response times.

Simply stated, a caching service applies the Pareto principle (aka the 80/20 rule) to *in-memory* data retrieval.[51] The Pareto principle is named after Vilfredo Pareto who, studying Italian economics, suggested in 1896 that 20% of the people of Italy owned approximately 80% of the land in Italy. This principle is widely and popularly exploited for focusing business management decisions and for fuelling continuous process improvement projects in many enterprises. It applies to IT as well, since application use cases commonly occur where a relatively small amount of the data is accessed most of the time (i.e., 20% of the data is accessed 80% of the time). Caching frequently accessed data directly in RAM -- as opposed to substantially slower retrieval from hard drives or SSDs -- offers the possibility of radically reduced response latency. Caching, in essence, means pay the one-time price up front and reap the benefits multiple times thereafter by obviating query re-execution.

Such is especially true for complex queries (e.g., queries requiring multi-table joins, hierarchical navigation, nested queries and subqueries, etc.) where initial query results are relatively slow. Caching the results of such long-running, expensive queries enables quick data retrieval -- possibly many times more rapidly -- especially if the application's data is somewhat static (i.e., where data changes are comparatively infrequent).

DAX enables the provisioning of a cluster of servers for implementing its caching service. Like Redshift nodes within a Redshift cluster, DAX cluster instances are also referred to as *nodes*. After a DAX cluster is launched within a VPC, the basic process describing how the DAX caching service is implemented can be summarized as follows:

- Assume that an application is deployed running on an EC2 instance endowed with *DAX client software*.
- The application executes DynamoDB API requests for accessing items from DynamoDB tables.
- The DAX client software intervenes -- acting as an intermediary -- and sends all DynamoDB API requests to the *endpoint* of the DAX cluster previously launched for the app. Similar to endpoints for previously described clusters, the DAX cluster endpoint is the front door for enabling load-balanced access to cluster nodes.
- By accessing data already cached in its memory (i.e., data previously retrieved from Dynamo tables), the DAX cluster determines whether or not it can satisfy API requests immediately, without accessing DynamoDB storage.
- If DAX finds the data in its cache then it processes the API request and directly replies to the application.
- If requested data is not found in cache, DAX tosses the request to DynamoDB and returns DynamoDB's resulting response back to the application.

Like previously describe database services, DAX is also a managed service and thus eases admin operational burdens. DAX supports the DynamoDB API and can be adapted for an existing application without substantial application modification.

Figure 7.4 positions DAX clusters relative to the overall super-type cluster hierarchy and further focuses on DAX entities and relationships.

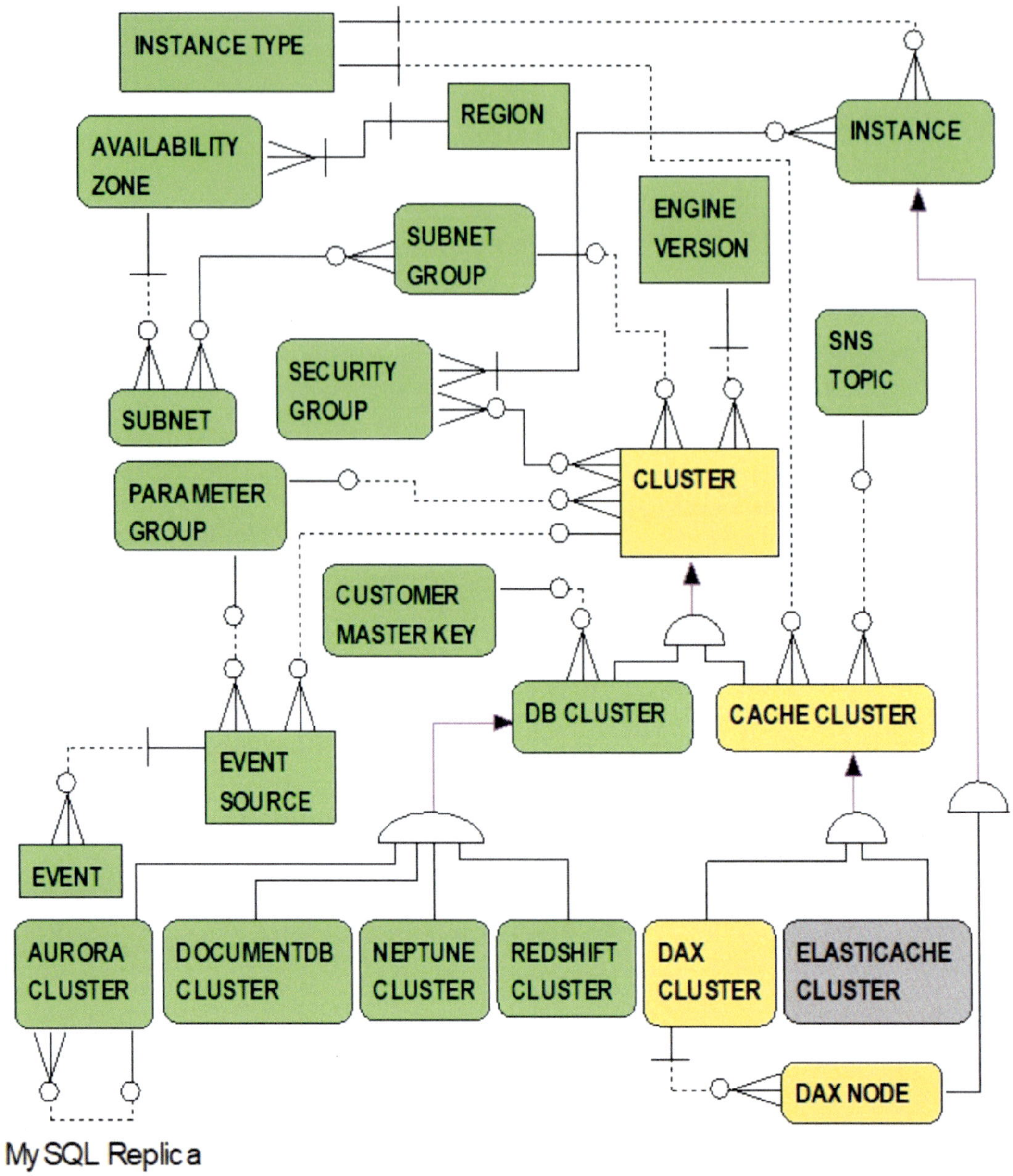

Figure 7.4 DynamoDB Accelerator (DAX) Entities and Relationships

From Figure 7.4, the CACHE CLUSTER entity represents the subset of all DAX and ElastiCashe clusters (ElastiCashe is addressed next in Chapter 8). The CACHE CLUSTER entity, in addition to inheriting higher level relationships to its super-type CLUSTER entity (i.e., many security groups, a single subnet group, a parameter group, an engine version, and serving as a source for event notifications) and CLUSTER attributes, provides additional relationships and attributes common to all cache clusters.

The one-to-many relationship from the INSTANCE TYPE entity to the CACHE CLUSTER entity indicates that each cache cluster is associated with a single instance type (i.e., all nodes in a cache cluster have the same node type). The one-to-many relationship from the SNS TOPIC entity to the CACHE CLUSTER entity identifies the SNS topic for publishing event notifications (e.g., email, text messages, etc.) triggered when cache cluster life events transpire. An SNS topic can apply to many cache clusters. Each cache cluster can be associated with a single SNS topic.

In addition to attributes inherited from its super-type CLUSTER entity, noteworthy sample attributes of the CACHE CLUSTER entity include *instance type* (foreign key), *sns topic* (foreign key), and *total nodes* (i.e., the total number of nodes in the cache cluster).

DAX Clusters

From Figure 7.4, the DAX CLUSTER entity represents the set of all DAX clusters. Note the entity name "DAX CLUSTER" and the term "DAX cluster" used herein correspond to the "*Cluster*" data type referred to in the Amazon DynamoDB API Reference. Because of the need (and greater good) for generalizing the cluster concept to a higher level, it was necessary to hijack the name "Cluster" (from the DynamoDB API documentation) and assign it to the CLUSTER entity so as to represent all clusters (database and cache clusters). Apologies are offered for any confusion this confiscation introduces. In the next chapter, a similar seizure is perpetrated for the concept of ElastiCache clusters.

Each DAX cluster is *a kind of* cache cluster and, therefore, directly inherits attributes and relationships from the CACHE CLUSTER entity, and indirectly inherits attributes and relationships from the CLUSTER entity. The one-to-many relationships from the INSTANCE TYPE and SNS TOPIC entities to the CACHE CLUSTER entity indicate that the DAX CLUSTER entity -- as a sub-type of the CACHE CLUSTER entity -- is indirectly associated with (i.e., inherits) a single instance type (i.e., all nodes in a DAX cluster have the same node type) and is also indirectly associated with (i.e., inherits) an SNS topic for communicating event notifications.

Common attributes inherited by the DAX CLUSTER entity from the CACHE CLUSTER entity include the sample attributes noted in the previous section (i.e., *instance type*, *sns topic*, and *total nodes*). Indirectly inherited from the CLUSTER entity, a DAX cluster can serve as a source for event notifications, and is associated with many security groups, a single subnet group, parameter group, and engine version. This means that a cluster of DAX nodes has:

- A set of security groups for controlling access to DAX cluster nodes.
- A subnet group within whose subnets DAX cluster nodes can be launched.
- A single parameter group specifying parameters for configuring all DAX cluster nodes.
- The DAX engine version for all cluster nodes.

Common attributes indirectly inherited by the DAX CLUSTER entity from the CLUSTER entity include the sample attributes noted in previous chapters which, for reader convenience, are replicated here as follows:

- *cluster identifier:* A user-supplied cluster identifier.
- *cluster create time:* The time when the cluster was created.
- *status:* Indicates the current state of the cluster.
- *endpoint:* Specifies the cluster's access information (i.e., DNS host name and port number) enabling applications to connect to the cluster.
- *preferred maintenance window:* Identifies the time period when maintenance activities (e.g., system upgrades) are preferred.

In addition to previously described attributes inherited from its super-type hierarchy (i.e., foreign keys and other attributes inherited from the CLUSTER and CACHE CLUSTER entities), noteworthy examples of attributes peculiar to the DAX CLUSTER entity include:

- *SSE status:* Maintains the DAX cluster's SSE (i.e., server-side encryption) status. Valid values include "ENABLED", "DISABLED", etc. When propagating in-memory data changes from the primary node to read replicas, DAX writes data to disk. DAX cluster data at-rest stored in the primary node and replica nodes can optionally be encrypted. DAX data cached in memory is not at-rest and not encrypted.
- *active nodes:* The total number of DAX cluster nodes that are operational and actively responding to requests.

DAX Nodes

Figure 7.4 also celebrates additional relationships peculiar to the DAX CLUSTER entity. The one-to-many relationship from DAX CLUSTER entity to the DAX NODE entity indicates that each DAX cluster controls one or more DAX node instances (i.e., all nodes running the same DAX cache engine version). The DAX NODE entity is also a sub-type of the INSTANCE entity indicating that a DAX node is *a kind of* EC2 instance. The functional purpose and intent of DAX nodes are sufficiently different justifying a distinct entity type name, hence the name "DAX NODE" entity (i.e., rather than attempting to generalize a new super-type for also encompassing the Redshift "RS NODE" entity name).

More specifically, like Aurora and Neptune (i.e., each with a primary DB instance and optionally many read replicas), a DAX cluster also controls a group of one or more nodes and uses similar vocabulary. A DAX cluster also has a *primary* node and, optionally, one or more *read replica* nodes. Unlike Redshift nodes (i.e., for exploiting parallel processing and data partitioning), the caching responsibilities for DAX nodes are substantially different.

When applications, using DAX client software, execute DynamoDB API *read operations* for accessing DynamoDB tables, such requests will be evenly dispersed by DAX among all DAX cluster nodes. Both the primary node and read replicas can process read operations. DAX nodes are responsible for satisfying these requests by checking if the requested data resides in cache memory with the hope of obviating access to DynamoDB storage. Specific table items in cache can be accessed via primary key values (i.e., maintained in an *Item Cache*). Beyond individual table items, larger *result sets* from previous table queries and scans are also accessible in cache memory (i.e., maintained in a *Query Cache*). For a cached result set in the Query Cache, the query's parameter values (i.e., rather than primary key values) are used for identifying the result set and enabling a *cache hit*.

In either case, if the specified table item or result set is discovered in cache, the requested data is directly sent to the requesting application by DAX. Slower access to DynamoDB storage is thereby avoided. A *cache miss* (i.e., not finding the requested data in cache memory), of course compels DAX to relay the request to DynamoDB. DynamoDB processes and returns either the item or a result set to DAX. DAX then responds to the application and appropriately updates either the item cache or query cache on the primary node.

The primary node is also responsible for dealing with data inserts and updates (i.e., *write operations*) to DynamoDB tables. Any changes to cached data on the primary node results in DAX synchronizing those changes to the read replica nodes if any. Read replicas never perform write operations to DynamoDB tables.

Note that the DAX default method for retrieving data is referred to as *eventually consistent reads*. This means that, because of the slight time lag for propagating changes to read replicas, it is entirely possible -- but with very low likelihood -- that the data DAX retrieves from cache might be out-of-date (i.e., inconsistent with most recent table updates). A quick follow-up read operation would most likely result in retrieving the most up-to-date data. Many applications can tolerate such behavior without risk of drastic consequences. For such use cases, eventually consistent reads are *close enough* for most practical purposes.

Recall that the concept of *eventually consistent reads* is in contrast with the previously described concept of *strongly consistent reads* which stringently ensure that all the latest data updates are included in the query result. For mission critical applications where possible retrieval of *stale* data is intolerable, DynamoDB optionally provides the ability to override the default *eventually consistent read* behavior in favor of *strongly consistent reads*. This option can be specified for individual read operations. For such read operations, cache access is bypassed guaranteeing direct access to freshest data in DynamoDB storage.

Similar to Aurora, Neptune and DocumentDB clusters, DAX read replicas also promote scalability and high availability. Horizontal scalability is supported by adding replicas to accommodate expanded concurrency requirements from a growing number of applications. Vertical scalability is supported by replacing the cluster with (i.e., forcing the creation of) a new cluster equipped with more potent node types for achieving greater throughput.

High availability is supported by DAX monitoring the health of the primary node and automatically failing over to a read replica if a primary node failure is detected. The read replica is then anointed as the new primary node. Provisioning read replicas in multiple availability zones of course also promotes high availability.

Chapter Summary

Figure 7.5 summarizes DynamoDB basic concepts.

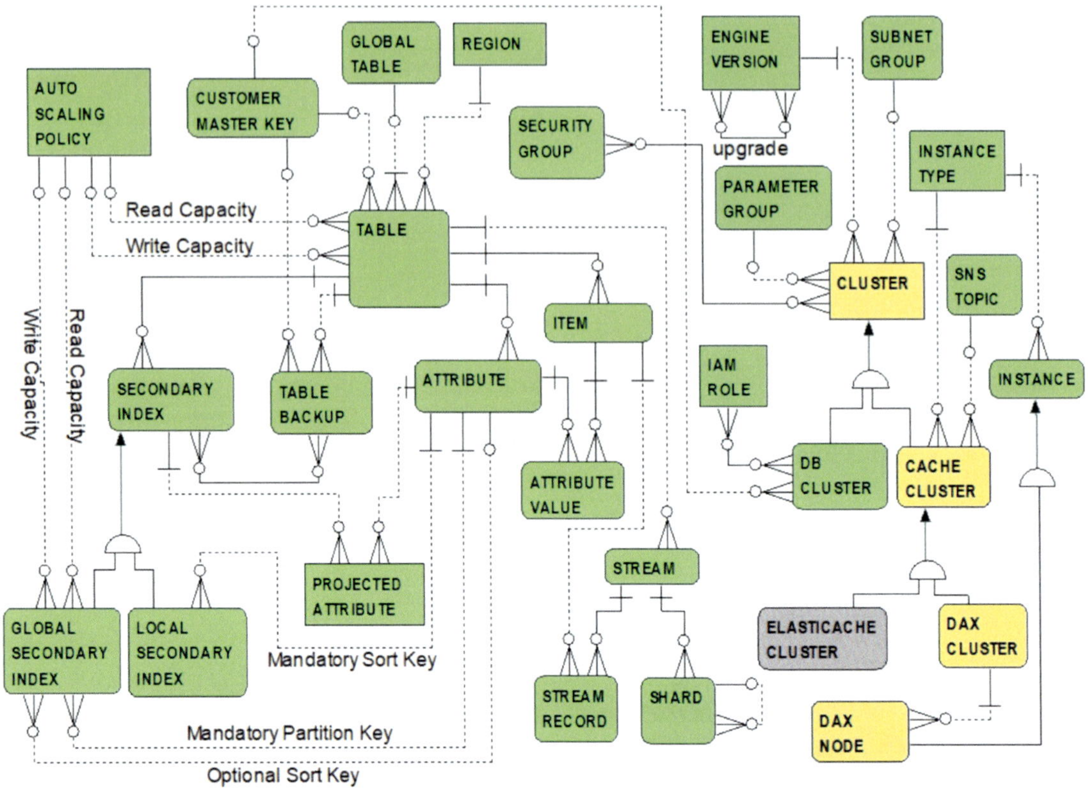

Figure 7.5 DynamoDB Entity Relationship Summary

Amazon DynamoDB -- like Neptune and DocumentDB -- is a managed NoSQL database service offered by AWS. DynamoDB supports both structured and semi-structured data based on key-value pairs and document storage (JSON documents). DynamoDB partitions data among SSDs and balances workload across server resources to satisfy latency and storage needs. DynamoDB replicates data across availability zones and automatically expands and repartitions storage without service interruption.

DynamoDB suits *internet scale* apps (e.g., e-commerce shopping carts) and is totally serverless, thereby obviating provisioning primary DB instances, replica DB instances, leader nodes or compute nodes as for previously described database services. Upfront capacity planning for new apps is simplified because DynamoDB can automatically scale resources to accommodate on-demand traffic. Capacity is specified in terms of *throughput requirements* instead of provisioning instances. Even though a NoSQL approach, DynamoDB supports *tables*. Throughput requirements are specified in terms of expected reads and writes on DynamoDB tables. DynamoDB transparently reserves sufficient servers and spreads data requests to accommodate anticipated throughput. Data partitions are also allocated to support throughput when additional storage is required. Separate *provisioned throughput* settings for *read capacity units* (RCUs) and *write capacity units* (WCUs) are specified for each table. If I/O traffic exceeds RCUs or WCUs, DynamoDB throttles and retries data requests without requiring special application logic.

Auto Scaling: DynamoDB exploits *auto scaling policies* (AWS Application Auto Scaling) for adjusting throughput settings based on a table's actual I/O traffic (i.e., *target utilization percentage* and upper/lower limits for RCUs/WCUs are specified).

Reserved Capacity: DynamoDB also provides cost saving opportunities by allowing for reserved capacity. However -- unlike RDS, Aurora, and Redshift where the purchase of reserved instances is offered -- DynamoDB offers the upfront purchase of RCUs and WCUs.

Tables and CRUD: DynamoDB's NoSQL approach supports a data manipulation language other than SQL for CRUD data operations -- but basic concepts for DynamoDB and the relational approach are comparable. Both include the definition of database *tables*. For RDBMS approaches, tables are defined in terms of *columns* whose values are maintained in table *rows*. DynamoDB tables are composed of *items* maintaining values of *attributes* (name-value pairs). Document data types (e.g., the "map" data type) support nesting collections of attributes (i.e., for supporting JSON files). DynamoDB tables are *schemaless* providing greater flexibility for defining tables and inserting items into a table. A PK (consisting of either one or two attributes) is required when defining DynamoDB tables. Other attributes and data types are not required to be defined when a DynamoDB table is created. Each item in a DynamoDB table can consist of its own set of different attributes and data types -- rather than an RDBMS, where columns apply to all rows and are normally defined when creating a relational table.

CRUD data operations for DynamoDB include: *PutItem* for adding an item to a table; *GetItem* for retrieving a single item by its primary key; *Query* enables retrieval of multiple items based on query filters; *UpdateItem* updates a single item; *DeleteItem* deletes one item at a time.

Global Tables: DynamoDB supports *global tables* for automatically replicating information across multiple regions. A global table is made up of a group of replica tables with the same table name, primary key, and AWS account -- but each replica table is located in a different region. Global tables essentially implement a *multimaster* database such that changes in any replica table are automatically propagated to other tables in the group. Global tables obviate app development effort and promote high availability, improved latency, world-wide customer access, disaster recovery, and business continuity.

Streams: DynamoDB provides *Streams* for tracking and capturing changes to table items -- aka change data capture (CDC) for inserts, updates, and deletions. Captured changes may then be used for triggering actions (e.g., sending email notifications for alerting events). A *stream record*, capturing a table item's primary key and "before" or "after" modifications, is stored for every modification to a table item. Stream records are clustered into groups known as *shards* which are automatically created and deleted as necessary by DynamoDB.

DynamoDB Accelerator (DAX): DAX is a *caching* service promoting rapid response times and cost savings for certain use cases. A caching service applies the Pareto principle (aka the 80/20 rule) to in-memory data retrieval (i.e., for apps where a relatively small amount of data is most frequently accessed). DAX enables provisioning a cluster of cache servers (aka cache cluster *nodes*). DAX assumes that an EC2 instance, running DAX client software, executes DynamoDB API data requests. DAX client software sends all data requests to the DAX cluster. If DAX finds the data in cache (i.e., a *cache hit* of data previously retrieved from Dynamo tables), then DAX directly replies to the application. If requested data is not found (a *cache miss*), DAX transfers the request to DynamoDB and returns DynamoDB's response to the application.

DAX Clusters: A DAX cluster is *a kind of* cache cluster which in turn is *a kind of* cluster. Because a DAX cluster is *a kind of* cache cluster, a DAX cluster inherits a single instance type (i.e., all nodes in a DAX cluster have the same node type) and an SNS topic for communicating event notifications. Because a cache cluster is *a kind of* cluster, a DAX cluster indirectly inherits security groups, a subnet group, a parameter group, an engine version, and is a source for event notifications.

DAX Nodes: A DAX cluster controls a group of one or more nodes and uses vocabulary similar to Aurora and Neptune (i.e., a *primary* node and possibly many *read replica* nodes). Unlike Redshift nodes (i.e., for MPP and data partitioning), the caching responsibilities for DAX nodes are substantially different. DAX balances DynamoDB API read operations among all DAX cluster nodes. Both the primary node and read replicas can process read operations. DAX maintains both an *Item Cache* (for table items accessed by PK) and a *Query Cache* (for *result sets* from previous queries -- accessed by query parameter values). The primary node performs *write operations* to DynamoDB tables. Any changes to cached data on the primary node are synchronized to read replica nodes.

By default, retrievals are *eventually consistent* (i.e., data retrieved from cache might be inconsistent with most recent table updates but close enough for most practical purposes). For mission critical apps intolerant of *stale* data, DynamoDB optionally overrides *eventually consistent read* behavior in favor of *strongly consistent reads* (i.e., cache access is bypassed with direct access to fresh data).

DAX read replicas promote scalability (horizontal and vertical) and high availability (automatically failing over to a read replica if a primary node fails).

Chapter 8 – Amazon ElastiCache

Amazon ElastiCache Introduction

Analogous to DAX, Amazon ElastiCache (ElastiCache) is another managed caching service that facilitates the implementation of scalable clusters of cache nodes. Caching services -- introduced in the previous chapter -- offer a phenomenal opportunity for improving response time latency by swiftly responding to queries with in-memory data and thereby minimizing an application's need to access data stored on slower hard disk drives or solid state storage devices. Also similar to DAX, ElastiCache -- as a managed service -- alleviates worries surrounding admin tasks, such as provisioning servers, installing and upgrading software, managing capacity changes, dealing with server or cluster failures, etc.

ElastiCache supports two general options for caching data in memory based on well-known, open-source software initiatives: the Memcached cache engine[52] and the Redis (**RE**mote **DI**ctionary **S**erver) cache engine.[53] Multiple engine versions are supported for each of these two options. High level pros and cons for comparing and contrasting Memcached and Redis cache engines are briefly summarized as follows:

- Both Memcached and Redis cache engines …

    - Deliver a NoSQL key-value store for maintaining and manipulating in-memory data as key-value pairs.[54]
    - Enable data partitioning across multiple nodes.
    - Promote high availability by allowing cluster nodes to be distributed among multiple availability zones.
    - Enable capacity adjustment in response to changes in demand by allowing cluster nodes to be appropriately added or removed.

- In addition to the above capabilities, the Redis cache engine offers a robust set of bonus options including:

    - Data encryption for data in-transit and data at-rest.
    - Replication of data across multiple cluster nodes, *read replicas,* for improving response times and high availability.
    - Failure detection and automatic failover (i.e., similar to DAX clusters, automatic failover of the *primary node* to a *read replica* node).
    - Creation of backups for recovery of existing clusters or populating data in new clusters.

The ElastiCache caching service is similar to the DAX caching approach -- as described in the previous chapter -- but is subtly different relating to how the application interfaces with ElastiCache and a back-end database (if any). The ElastiCache caching process is briefly summarized in terms of using a *lazy loading* caching strategy as follows:[55]

- Assume that an application equipped with appropriate Memcached or Redis client API software is deployed running on an EC2 instance.
- Assume that the application, in addition to exploiting ElastiCache, also requires a back-end database.
- When retrieving data, the application initially requests data from an ElastiCache cluster via either Redis or Memcached API commands.
- If a cache hit occurs, ElastiCache returns the data to the application, thereby freeing the application from having to access the back-end database.
- If a cache miss occurs (i.e., requested data is not found or expired in cache), the application then follows up with an additional request for directly accessing the database.
- The application then issues appropriate API commands for writing the fresh data to ElastiCache so as to facilitate future retrievals (i.e., cache hits).

Recall from the previous chapter that the DAX client intercepts an application's DynamoDB API requests and transparently acts (i.e., if a DAX cache miss occurs) as an intermediary to DynamoDB storage. For DAX, the application issues DynamoDB API requests and is oblivious to DAX caching activity. Application code works the same with or without DAX. But ElastiCache is different. An application using ElastiCache issues Memcached or Redis commands for accessing cached data and, independently, might also issue separate DML statements for accessing a possible back-end database. ElastiCache processes the command from the application and, unlike DAX, is oblivious of any back-end database. ElastiCache interfaces only with the application. If a cache miss occurs, ElastiCache does not come to the rescue by retrieving the data, like DAX, from a back-end database

Note that the lazy load strategy caches only requested data (i.e., as a result of a cache miss) and can be implemented with a *time to live* (i.e., TTL, an integer value) indicator specifying the number of seconds after which the key *expires*.[56] Any attempt to access expired data in cache is considered to be a cache miss resulting in potentially *stale* data being refreshed from the database -- and of course resetting the TTL.

An alternative or supplement to the lazy loading strategy is known as the *write through* caching strategy.[57] In this case, whenever the application inserts or updates data in its database, it also updates the cache. Cached data, therefore, is always current and consistent with a back-end database.

The Redis and Memcached data manipulation approaches for creating, reading, updating and deleting in-memory data substantially diverge from previously described DBMS engines. The architecture and design details of the Memcached and Redis cache engines are well beyond the intent and scope of this immediate conceptual data modeling endeavor for graphically summarizing fundamental ElastiCache concepts. Fortunately however, abundantly detailed ElastiCache documentation is readily and freely available (i.e., separate user guides for Memcached and Redis cache engines, ElastiCache API documentation, and AWS CLI Command Reference for ElastiCache).[58] These documents provide plentiful information -- and additional references -- allowing for fully evaluating the capabilities, features, functions, benefits and limitations of each of these alternatives. Specific application use cases are also identified where one caching approach might be considerably more enticing than the other. The following sections respectively introduce specific Redis and Memcached concepts helpful for understanding ElastiCache data model diagrams about to be revealed.

Redis Introduction

The primary focus of Redis is to always access and manipulate data within a cache server's random access memory (RAM). For some use cases, Redis obviates, or at least diminishes, the need for persisting data in a back-end database. This focus epitomizes a substantial shift from previously described DMBS approaches which emphasize maintaining data on secondary storage devices (i.e., HDDs or SSDs). The Redis approach contrasts with DAX where cached in-memory data acts as an intermediary to data stored in a DynamoDB database. Redis combines both an in-memory cache and, secondarily, allows for storing data on disk for persisting in-memory data. But Redis data stored on disk is not directly accessed or manipulated. Redis CRUD operations are entirely performed in-memory. Redis in-memory data can be backed up in snapshots. Changes to Redis in-memory data can also be written to log files. Snapshots and log files serve as the basis for recovering in-memory data to support restarting servers and automatic failover from the primary node to a read replica. Confidence in automatic failover, high availability, and disaster recovery capabilities render (for some applications) a backend database as an afterthought and possibly not a fundamental requirement.

In addition to its in-memory focus, the Redis data manipulation language (i.e., DML) is by itself yet another major shift from the DML paradigms of previously described RDBMS and NoSQL approaches. The major difference is that Redis DML commands do not neatly map to previously noted SQL or NoSQL CRUD commands (e.g., for accessing relational tables, graph vertices, documents, etc.). Instead, Redis deals with higher level *abstract data types* which transcend fundamental DBMS table and column concepts. Redis exceeds the expectations of a conventional key-value store and is, therefore, also described as a data structures server.[59] Redis provides a set of commands specific to each of its supported abstract data types.

Examples of such higher-level Redis data structures, concepts and associated commands include the following:[60]

- **Strings**: A Redis String is simply a sequence of bytes and serves as the foundation for other Redis data structures. Akin to a blob data type (supported by RDBMs such as Oracle DB and IBM Db2),[61] a Redis string contains binary-safe data (i.e., any kind of data either binary or text data) up to maximum of 512 megabytes.[62] For example, a Redis key is itself a string whose value can be mapped to another string. A simple use case is a string holding the result set of a SQL query from a relational database. This result set (i.e., a character string) is then mapped to a Redis key whose string value is the actual text of the SQL query. This key-value pair can then be cached for later retrieval.

  Redis provides many commands for accessing and manipulating strings including commands for setting and getting string and substring values, incrementing and decrementing integer and floating values, appending bytes to the end of string, and setting the TTL expiration of a key. Specific examples of String related commands include:

- **SET, GET**: Separate commands for respectively assigning a string value and fetching the string value for a specified key. For the SET command, the Redis command line interface provided by the Redis client accepts characters for setting a string's value. A programming language API is typically used for moving binary data to a string. To use the Redis CLI SET command for binary data, the binary data must first be encoded (i.e., serialized using a Base64 encoding standard).[63] Encoded binary data is acceptable input for the SET command.
- **SETRANGE, GETRANGE**: Separate commands for respectively setting a substring value and fetching a substring value for a specified key.
- **MSET, MGET**: Separate commands for respectively setting string values and fetching string values for multiple keys.
- **INCRBY**, **DECRBY**: Separate commands for respectively incrementing and decrementing an integer value by a specified number.
- **APPEND**: Attaches (i.e., concatenates) a value to the current string value of a specified key.
- **SETEX**: Specifies a value for setting the TTL expiration of a key.

- **Lists**: A List is an ordered sequence of string values. Values can be duplicated within a List. The elements are linked according to their order of insertion. Examples of List related commands include:

  - **LPUSH**: Inserts new elements at the head of a list (i.e., at the left)
  - **RPUSH**: Inserts new elements at the end of a list (i.e., at the right)
  - **LRANGE**: Retrieves a range of elements from a list by specifying the first and last elements to be extracted.

- **Sets**: Comparable to a List, a Set is an unordered series of strings. The values are non-repeating and are in no specific order. Examples of Set related commands include:

  - **SADD**: Inserts new elements to a Set.
  - **SUNION**: Returns a Set whose elements are the union between multiple Sets.
  - **SISMEMBER**: Determines if an element is a member of the set by returning either a 1 or 0 (i.e., respectively indicating true or false).

- **Sorted Sets**: Akin to an ordinary Set, a sorted set is a series of unique values but each value is linked to a *score* -- a floating point number -- which can be used for sorting the set. Unlike an ordinary Set -- but similar to a List -- Sorted Sets enable retrieving a range of elements. Examples of Sorted Set related commands include:

  - **ZADD**: Similar to SADD for ordinary Sets, ZADD inserts new elements along with their associated score to a Sorted Set.
  - **ZRANGE**: Similar to LRANGE for Lists, ZRANGE retrieves a range of elements from a Sorted Set -- sorted from highest to lowest score -- by specifying the first and last elements to be extracted.
  - **ZREVRANGE**: Similar to ZRANGE, but ZREVRANGE **rev**erses the sort order.

- **Hashes**: Comparable to a row in a relational database, a Hash is simply a mapping of one or more key-value pairs (i.e., string fields mapped to string values) associated with a unique *hash key*. A Hash essentially represents the data structure for any specific object of interest to an application. Redis Hash commands allow access or updates to a single field or multiple fields at a time. Examples of Hash related commands include:

    - **HMSET**: Sets multiple fields of the Hash for a specified hash key field.
    - **HMGET**: Returns an array of values for a specified hash key field and specified field names.
    - **HEXISTS**: For a specified hash key field, HEXISTS verifies that a specified hash field exists.

- **Pub/Sub**: Beyond the above samples of supported abstract data types, Redis also facilitates the implementation of a variety of application capabilities such as pub/sub services for **pub**lishing messages to *channel* **sub**scribers. Publishers (i.e., message senders) post messages to a channel, which links to message receivers who previously subscribed to the channel. Examples of Pub/Sub related commands include:

    - **SUBSCRIBE**: Enables a receiver to subscribe to (i.e., listen for messages from) one or more specified channels.
    - **PUBLISH**: Enables a publisher to post a message to a specified channel.
    - **UNSUBSCRIBE**: Discontinues the receipt of messages from one or more specified channels.

Memcached Introduction

Memcached is less robust compared to Redis especially in terms of supported data structures. Unlike Redis, Memcached is not a professed data structure server. Lists, Sets, Sorted Sets, Hashes, and Pub/Sub services are not supported. Memcached simply provides an ordinary, conventional key-value store where strings are the only supported data type. Memcached essentially enables a large in-memory hash table. Simply stated, a key string value is mapped to another string value. A Memcached string, like a Redis string, is a block of raw data consisting of a sequence of bytes. The string size can be configured up to a maximum of 128 megabytes, substantially smaller than Redis strings, but more than adequate for enabling solutions satisfying requirements for many business-critical caching use cases. Unlike Redis, Memcached strings are not binary-safe. Binary data cannot be directly assigned to a Memcached string value without encoding. As previously mentioned for setting Redis string values using the Redis SET CLI command, values moved to Memcached strings must be encoded (e.g., pre-serialized using Base64).

Like Redis, Memcached APIs for manipulating string values are supported by many familiar, popular programming languages. Examples of Memcached string related commands include:

- **Set Data**: Specifies a new or existing key and maps a new value to the specified key (i.e., overwriting the value for an existing key).

- **Add Data**: Adds a new key and sets its value but does not overwrite the value for an existing key. The command fails if the specified key already exists.
- **Replace Data**: Specifies an existing key and replaces its current value. The command fails if the specified key does not exist.
- **Append Data**: Specifies an existing key and appends (i.e., attaches) data to the end (i.e., after the last byte) of the current value. The command fails if the specified key does not exist.
- **Prepend Data**: The opposite of Append Data, this inserts data in front of the current value of an existing key.
- **Get Data**: This is used to access the value stored at one or more specified keys. If a specified key does not exist, nothing is returned.
- **Delete Key**: This is used to remove an existing key-value pair from the cache.
- **Incr:** For strings with integer values, this increments the numeric value of an existing key by a specified positive value.
- **Decr**: For strings with integer values, this decrements the numeric value of an existing key by a specified positive value.

ElastiCache Clusters

Figure 8.1 portrays where ElastiCache clusters fit relative to the overall super-type cluster hierarchy and further focuses on entities and relationships specific to ElastiCache.

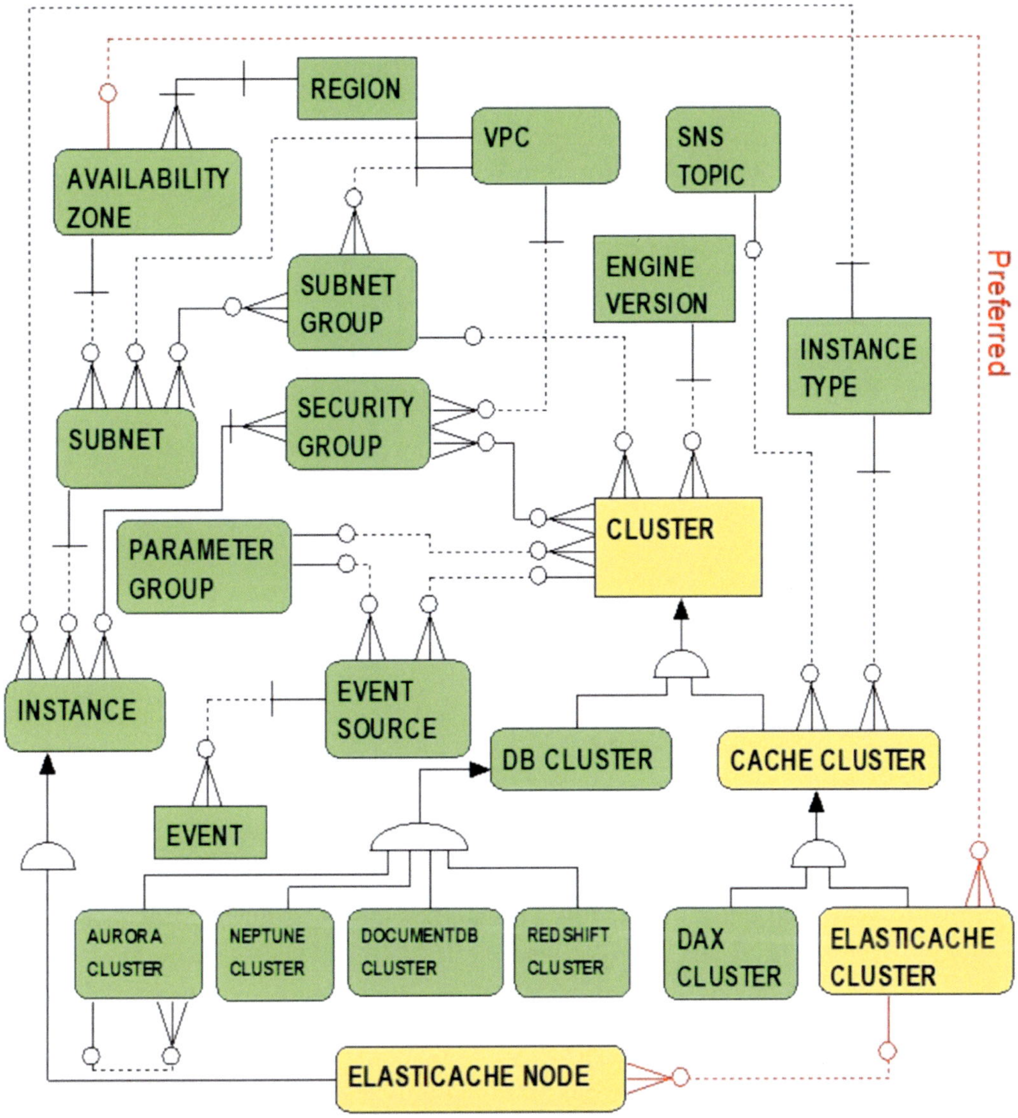

Figure 8.1 Entities and Relationships … ElastiCache Clusters

From Figure 8.1, the CACHE CLUSTER entity (as previously noted in Chapter 7) represents the subset of all cache clusters, both DAX and ElastiCache clusters. The ELASTICASHE CLUSTER entity is *a kind of* cache cluster (i.e., a sub-type of the CACHE CLUSTER entity) and, therefore, directly inherits relationships and attributes from its parent CACHE CLUSTER super-type entity and indirectly inherits relationships and attributes from the its grandparent CLUSTER super-type entity. Similar to the name clash described previously when dubbing the DAX CLUSTER entity (i.e., recall that a DAX cluster is simply referred to as a "Cluster" in the DynamoDB API documentation), note also that the entity name "ELASTICASHE CLUSTER" and the term "ElastiCache cluster", used herein, correspond to the "Cache Cluster" data type referred to in the ElastiCache API documentation (i.e., the Amazon ElastiCache API Reference).

As previously noted, because of the need -- and greater good -- for generalizing the cache cluster concept to a higher level, it was necessary to hijack the name "Cache Cluster" for the super-type CACHE CLUSTER entity so as to represent all cache clusters (i.e., ElastiCache and DAX cache clusters) not just ElastiCache clusters. Apologies are again offered for any confusion created by this additional name confiscation.

The CACHE CLUSTER entity, in addition to inheriting higher level relationships to its super-type CLUSTER entity (i.e., many security groups, a single subnet group, a parameter group, an engine version, and a source for event notifications) and CLUSTER attributes, provides additional relationships and attributes common to all cache clusters in general. The one-to-many relationships from the INSTANCE TYPE entity and the SNS TOPIC entity to the CACHE CLUSTER entity imply that the ELASTICASHE CLUSTER entity -- as a sub-type of the CACHE CLUSTER entity -- is indirectly associated with (i.e., inherits) a single instance type (i.e., all nodes in an ElastiCashe cluster have the same node type) and is also indirectly associated with (i.e., inherits) an SNS topic for communicating event notifications (e.g. email, text messages, etc.) triggered when ElastiCashe cluster life events transpire.

Common attributes inherited by the ELASTICASHE CLUSTER entity from the CACHE CLUSTER entity include the sample attributes noted in the previous chapter: *instance type* (foreign key), *sns topic* (foreign key), and *total nodes* (i.e., the total number of nodes in the ElastiCashe cluster).

Indirectly inherited from the CLUSTER entity, an ElastiCashe cluster can serve as a source for event notifications, and is associated with many security groups, a single subnet group, parameter group, and engine version. This means that a cluster of ElastiCashe nodes has:

- A set of security groups for controlling access to ElastiCashe cluster nodes.
- A subnet group within whose subnets ElastiCashe cluster nodes can be launched.
- A single parameter group specifying parameters for configuring all ElastiCashe cluster nodes.
- The ElastiCashe engine version for all cluster nodes.

Common attributes indirectly inherited by the ELASTICASHE CLUSTER entity from the CLUSTER entity include the sample attributes noted in previous chapters which, for convenience, are replicated here as follows:

- *cluster identifier:* A user-supplied cluster identifier.
- *cluster create time:* The time when the cluster was created.
- *status:* Indicates the current state of the cluster.
- *endpoint:* Specifies the cluster's access information (i.e., DNS host name and port number) enabling applications to connect to the cluster.
- *preferred maintenance window:* Identifies the time period when maintenance activities (e.g., system upgrades) are preferred.

In addition to the above attributes inherited from its super-type hierarchy, noteworthy examples of attributes peculiar to the ELASTICACHE CLUSTER entity include:

- *at rest encryption enabled*: Indicates whether or not encryption is enabled for data at-rest.
- *transit encryption enabled*: Indicates whether or not encryption is enabled for data in-transit.

- *snapshot retention limit*: For automatically created cluster snapshots, this indicates the number of days such backups are preserved before automatically being deleted. A value = 0 indicates that automatic backups are disabled.
- *snapshot window*: Specifies the daily time range when cluster snapshots are created.

Figure 8.1 also highlights additional relationships peculiar to the ELASTICACHE CLUSTER entity. The one-to-many relationship from AVAILABILITY ZONE entity to the ELASTICACHE CLUSTER entity indicates that each ElastiCache cluster can be associated with a single *preferred availability zone* in which cluster nodes would rather be located (i.e., when nodes are not to be launched in multiple availability zones).

The one-to-many relationship from the ELASTICACHE CLUSTER entity to the ELASTICACHE NODE entity indicates that each ElastiCache cluster controls one or more ElastiCache node instances. However, the rules underlying this one-to-many relationship are subtly different depending on whether the ElastiCache cluster is a Memcached cluster or a Redis cluster.

For Memcached, an ElastiCache cluster controls one to a maximum of 20 ElastiCache nodes. If two or more ElastiCache nodes are provisioned for a Memcached cache engine, data is partitioned among those nodes. Partitioning means distributing cached data among multiple Memcached nodes such that each node contains a subset of key-value pairs. For a Memcached cluster, the concepts of primary node and read replica nodes do not apply. Memcached nodes are each independent of one another such that each node is oblivious to which key-value pairs are stored in its memory or the memory of other cache nodes. Memcached nodes are functionally equivalent. A hashing algorithm chooses the target node for retrieving, storing, or modifying a key-value pair. Why again is partitioning good? To facilitate and enhance scaling, performance and availability.

But for Redis, an ElastiCache cluster (i.e., within the scope of this one-to-many relationship) can have only <u>one node</u> -- not many nodes. This circumstance is referred to as a *Redis standalone cluster* (i.e., a cluster with only a single ElastiCache node running the Redis client). So in general, for emphasis, an ElastiCache cluster can be composed of one or more ElastiCache nodes, but for a *Redis standalone cluster* there is only one node. In other circumstances, an ElastiCache cluster for Redis can indeed control many more than 20 nodes, but the conceptual data model to account for such requires extensive enhancement, which is deferred as the subject of the next section, for introducing the concept of *Replication Groups*.

Note in Figure 8.1 that the ELASTICACHE NODE entity is also a sub-type of the INSTANCE entity indicating that an ElastiCache node is *a kind of* EC2 instance.

Figure 8.2 enhances 8.1 by appropriately annotating the one-to-many relationship from the ELASTICACHE CLUSTER entity to the ELASTICACHE NODE entity to emphasize the distinction between Memcached clusters and Redis standalone clusters.

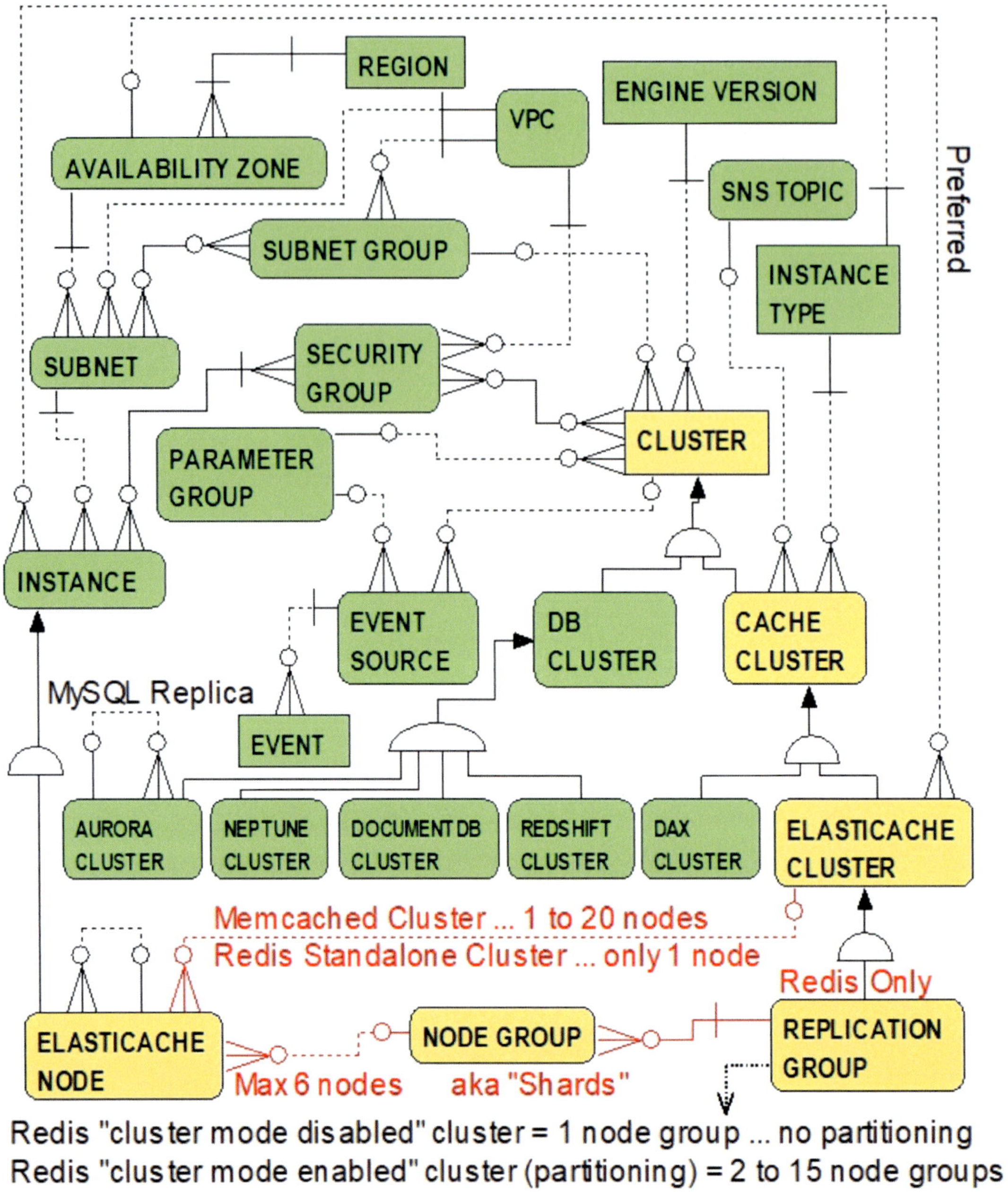

Figure 8.2 Memcache vs. Redis …. One-to-Many ElastiCache Nodes

From Figure 8.2, the one-to-many relationship from the ELASTICACHE CLUSTER entity to the ELASTICACHE NODE entity is annotated, above the relationship line, to reflect that a Memcached ElastiCache cluster controls 1 to a maximum of 20 Memcached nodes. The annotation below the relationship line reflects that a *Redis standalone cluster* can control only one Redis node.

ElastiCache Replication Groups

Figure 8.2 also introduces the concept of ElastiCache Replication Groups and the REPLICATION GROUP entity. Essentially, a replication group is an ElastiCache cluster for Redis which enables the provisioning of multiple Redis nodes. Replication groups also deliver options for replicating in-memory data and partitioning data across Redis nodes.[64] The REPLICATION GROUP entity is shown as a sub-type of the ELASTICACHE CLUSTER entity. From a data modeling perspective, each replication group is a special case of an ElastiCashe cluster (i.e., a replication group is *a kind of* ElastiCashe cluster). To emphasize, a replication group is an ElastiCashe cluster but an ElastiCashe cluster (e.g., a Memcached cluster or Redis standalone cluster as described in the previous section) is not necessarily a replication group.

So, why are replication groups special ElastiCashe clusters? Because there are special relationships and attributes that apply specifically to the REPLICATION GROUP entity which do not apply to its parent ELASTICACHE CLUSTER entity. More specifically, a replication group clusters Redis nodes into one or more *node groups* (aka *shards* ... not to be confused with DAX stream "shards" and the SHARD entity introduced in the previous chapter) to facilitate data replication and partitioning.

The one-to-many relationship from the ELASTICACHE CLUSTER entity to the NODE GROUP entity emphasizes that each replication group can be made up of one or more node groups. The one-to-many relationship from the NODE GROUP entity to the ELASTICACHE NODE entity emphasizes that each node group consists of one or more ElastiCache nodes. Note in Figure 8.2 that this relationship is annotated to reflect that each node group consists of a maximum of 6 nodes. Unlike Memcached, Redis node groups re-introduce the concepts of *primary nodes* and *read replica* nodes. One node, the primary node, is the one and only node in the node group used for both reading and writing data. In addition to the primary node, up to 5 nodes can be established in a node group as read replica nodes. Data written to the primary node is cloned asynchronously to potentially several replica nodes (for enhanced scalability) by enabling data retrievals distributed among several nodes. The one-to-many reflexive relationship from the ELASTICACHE NODE entity to itself indicates that the primary node of a Redis node group can have one or more ElastiCache nodes as read replicas.

Figure 8.2 is also annotated to emphasize special ElastiCache jargon to describe a replication group consisting of a single node group. Such is referred to as a *Redis (cluster mode disabled) cluster*. In this case, a single node group supports replication but disables data partitioning which requires at least 2 node groups. A replication group consisting of at least 2 node groups -- thereby enabling partitioning -- is referred to as a *Redis (cluster mode enabled) cluster* and can be partitioned up to a maximum of 15 node groups. In either case, a node group with a primary node and at least one replica node enables automatic failover to a replica node in the event of primary node failure. The replica node is promoted to become the primary node. ElastiCache refers to this as *Multi-AZ with automatic failover*.

Note that the terminology used in the ElastiCache console emphasizes the term *cluster* rather than *replication group* for all of the above cases. The ElastiCache console uses the term *cluster* for Redis (cluster mode enabled) clusters, Redis (cluster mode disabled) clusters, Redis standalone clusters, and Memcached clusters. The terminology emphasized in ElastiCache API and CLI documentation reserves the term *cluster* for Memcached and Redis standalone clusters (e.g., the CreateCacheCluster API action), and uses the term *replication group* for Redis (cluster mode enabled) clusters and Redis (cluster mode disabled) clusters (e.g., the CreateReplicationGroup API action). Respectively using the names ELASTICACHE CLUSTER and REPLICATION GROUP for the entity names in Figure 8.2 is an attempt at maintaining compatibility within the data model and consistency with vocabulary nuances in ElastiCache documentation.

ElastiCache Backups

Figure 8.3 introduces the concept of ElastiCache snapshots.

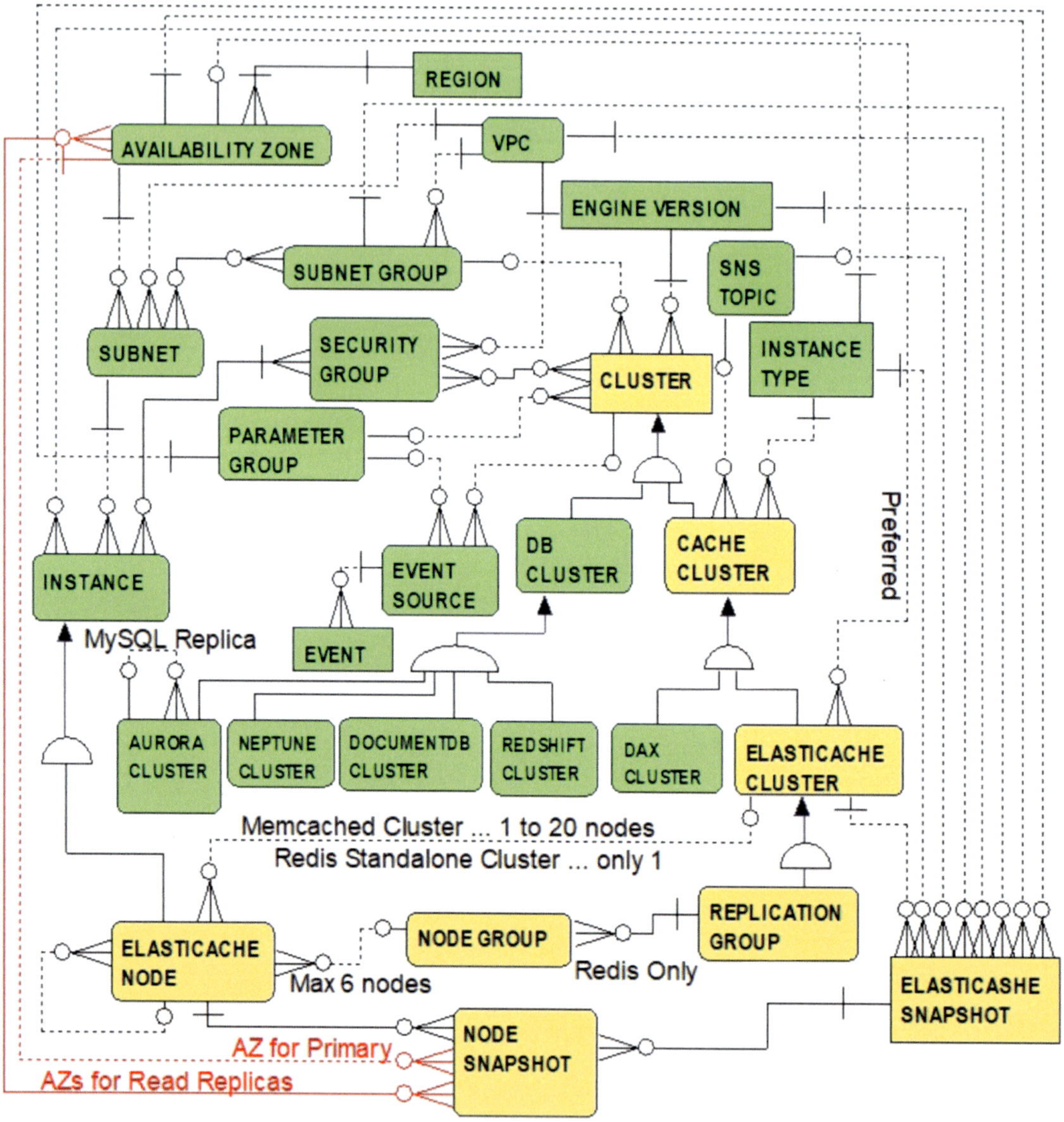

Figure 8.3 Entities and Relationships … ElastiCache Backups

From Figure 8.3, the one-to-many relationship from the ELASTICACHE CLUSTER entity to the ELASTICACHE SNAPSHOT entity indicates that each ElastiCache cluster (e.g., all primary nodes provisioned for a Redis ElastiCache cluster) can be backed up by one or more ElastiCache *snapshots*. Backups in this context apply only to Redis clusters. The ELASTICACHE SNAPSHOT entity (similar to the DB CLUSTER SNAPSHOT entity previously introduced in Chapter 3) duplicates crucial relationships (i.e., for the ELASTICACHE CLUSTER, CACHE CLUSTER, and CLUSTER entities) in order to fulfil its destiny to accurately restore an ElastiCache cluster if and when necessary. These duplicated relationships include one-to-many relationships from INSTANCE TYPE, SNS

TOPIC, ENGINE VERSION, VPC, SUBNET GROUP, AVAILABILITY ZONE, and PARAMETER GROUP entities to the ELASTICACHE SNAPSHOT entity. These relationships reflect the essential state of the Redis ElastiCache cluster at the time of the backup (i.e., the instance type, SNS topic, engine version, VPC, subnet group, availability zone, and parameter group in effect at the time of snapshot creation).

In addition to attributes derived from the above noted one-to-many relationships to the ELASTICACHE SNAPSHOT entity (i.e., foreign keys), noteworthy examples of additional attributes of the ELASTICACHE SNAPSHOT entity include:

- *snapshot name:* The snapshot name either automatically generated or chosen by the user (i.e., for a *manual snapshot*).
- *snapshot retention limit:* The number of days that a snapshot is saved after which it is removed from storage (i.e., for automatically created snapshots). Manual snapshots are not removed automatically and must be removed explicitly (e.g., via a *DeleteSnapshot* API call). For manual snapshots, *snapshot retention limit* is the value that was stored for the backed-up cluster and is otherwise disregarded.
- *snapshot source:* Indicates a value of either "automated" or "manual" respectively reflecting whether the snapshot was automatically generated or requested by the user.
- *snapshot status:* Reflects the snapshot status having valid values such as "creating," "available," "deleting," etc.

The NODE SNAPSHOT entity in Figure 8.3 resolves a many-to-many relationship between the ELASTICACHE NODE and ELASTICACHE SNAPSHOT entities. Specifically, an ElastiCache node can be included on many snapshots and, clearly, a snapshot can backup many ElastiCache nodes. The NODE SNAPSHOT entity resolves this relationship by reflecting the *node group configuration* information (e.g., availability zones) for each node included in the snapshot. Indeed, only one node from each node group actually needs to be backed up. The one-to-many relationship from the ELASTICACHE NODE entity to the NODE SNAPSHOT entity reflects that each node snapshot must be associated with a single ElastiCache node. For a Redis (cluster mode enabled) cluster, the primary node of each node group is backed up. For a Redis (cluster mode disabled) cluster, any node within the single node group can be selected for use during the backup process. This allows the option of backing up a read replica instead of the primary node and thereby enabling the primary node to respond to requests unperturbed by the backup process.

The one-to-many relationship from the AVAILABILITY ZONE entity to the NODE SNAPSHOT entity identifies the availability zone for the node group's primary node at the time the snapshot was taken. The many-to-many relationship between the AVAILABILITY ZONE entity and NODE SNAPSHOT entity identifies the availability zones for each replica node within the node group.

In addition to attributes derived from the above noted one-to-many relationships to the NODE SNAPSHOT entity (i.e., foreign keys), noteworthy examples of additional attributes of the NODE SNAPSHOT entity include:

- *snapshot create time*: The date and time at which point the node's cache data was accessed for creating the backup.

- *cache node create time:* The date and time at which point the node itself was created within the cluster for which the snapshot was created.
- *node group id:* The unique identifier (i.e., either the user supplied id or the Redis supplied id) for the node group in which the backed-up node is included.
- *slots:* For partitioning data across node groups, this specifies a range of values between 0 and 16383 (e.g., "8000-12000") for the *keyspace* associated with the node group. A total of 16384 hash slots are used by the Redis hashing algorithm in determining the slot for a specified key (i.e., a CRC16 hashing algorithm). If this slot range is unspecified, key spaces are equally distributed by ElastiCache between the node groups.

ElastiCache Reserved Instances, Node Type Parameters

ElastiCache (like RDS, Aurora and Redshift) enables purchasing reserved instances for cache nodes thereby offering considerable cost savings compared to on-demand pricing. Figure 8.4 extends Figure 8.3 by emphasizing that reserved instances and the concept of *node type parameters* also apply to ElastiCache clusters.

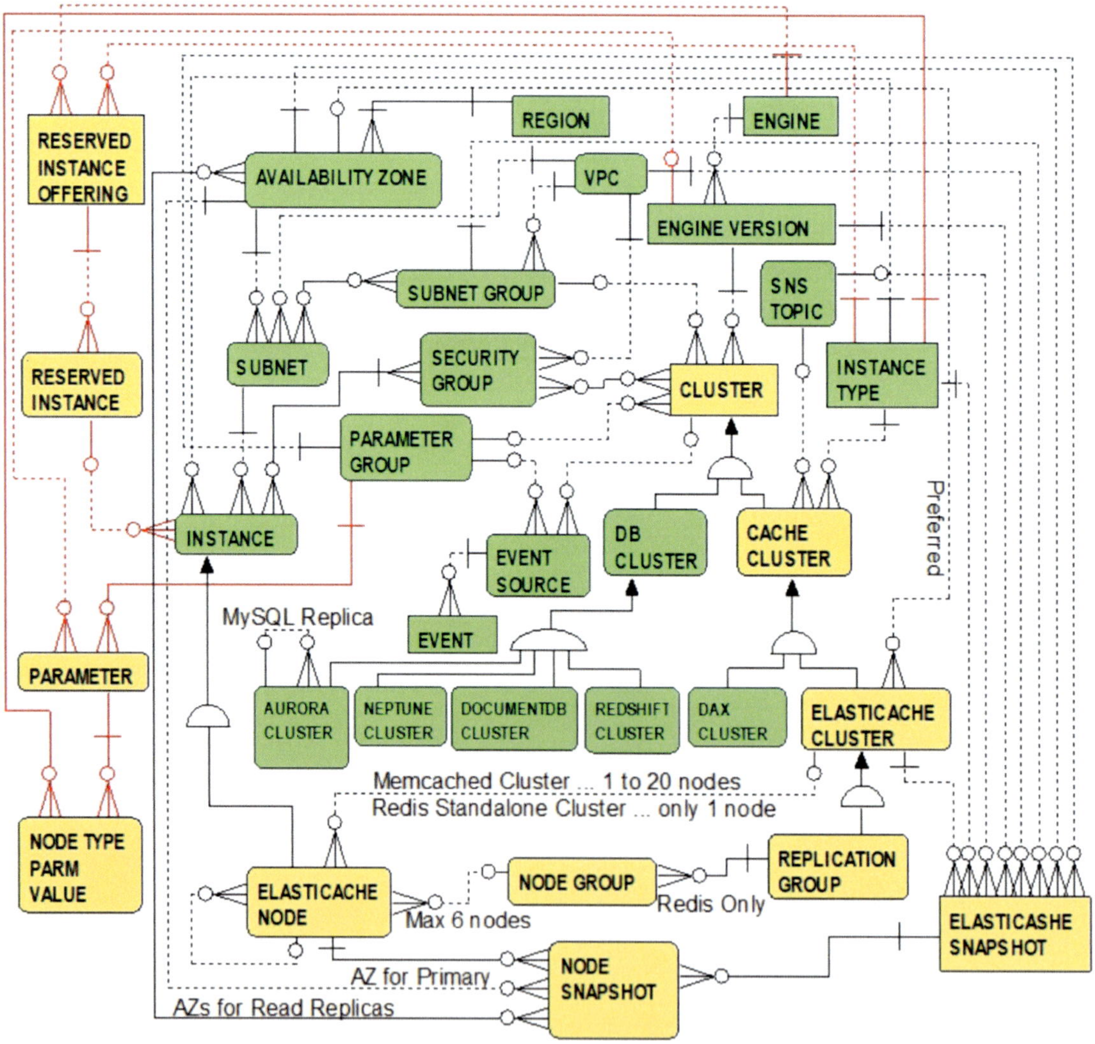

Figure 8.4 Reserved Instances, Node Type Parameters

From Figure 8.4, the RESERVED INSTANCE OFFERING entity and RESERVED INSTANCE entity are re-introduced (from Chapter 2, Figure 2.4) highlighting reserved instance fundamental concepts. Recall that the RESERVED INSTANCE OFFERING entity represents the starting point when a user searches for discount offerings prior to actually purchasing reserved instances. The one-to-many relationship from the INSTANCE TYPE entity to the RESERVED INSTANCE OFFERING entity indicates that each reserved instance offering applies to a single instance type (i.e., a cache node type). Each reserved instance offering can spawn one or more reserved instances, as represented by the one-to-many relationship between the RESERVED INSTANCE OFFERING entity and the RESERVED INSTANCE entity. Also recall that the one-to-many relationship between the RESERVED INSTANCE and INSTANCE entities suggests that each reserved instance (i.e., the discounted pricing associated with each reserved instance) can apply to one or more actually running cache nodes.

Note also the NODE TYPE PARM VALUE entity in Figure 8.4. This resolves a many-to-many relationship between the INSTANCE TYPE and PARAMETER entities reflecting that specific values of parameters can differ depending on the cache node type. Attributes of the NODE TYPE PARM VALUE entity include *node type*, *parameter name*, and the parameter *value* specific to the node type. Each cache node type can have many parameters for which special values apply. Examples of Memcached parameter names whose values vary depending on the cache node type include max cache memory (i.e., *max_cache_memory)* and number of threads (i.e., *num_threads).*

Chapter Summary

Figure 8.5 summarizes ElastiCache basic concepts.

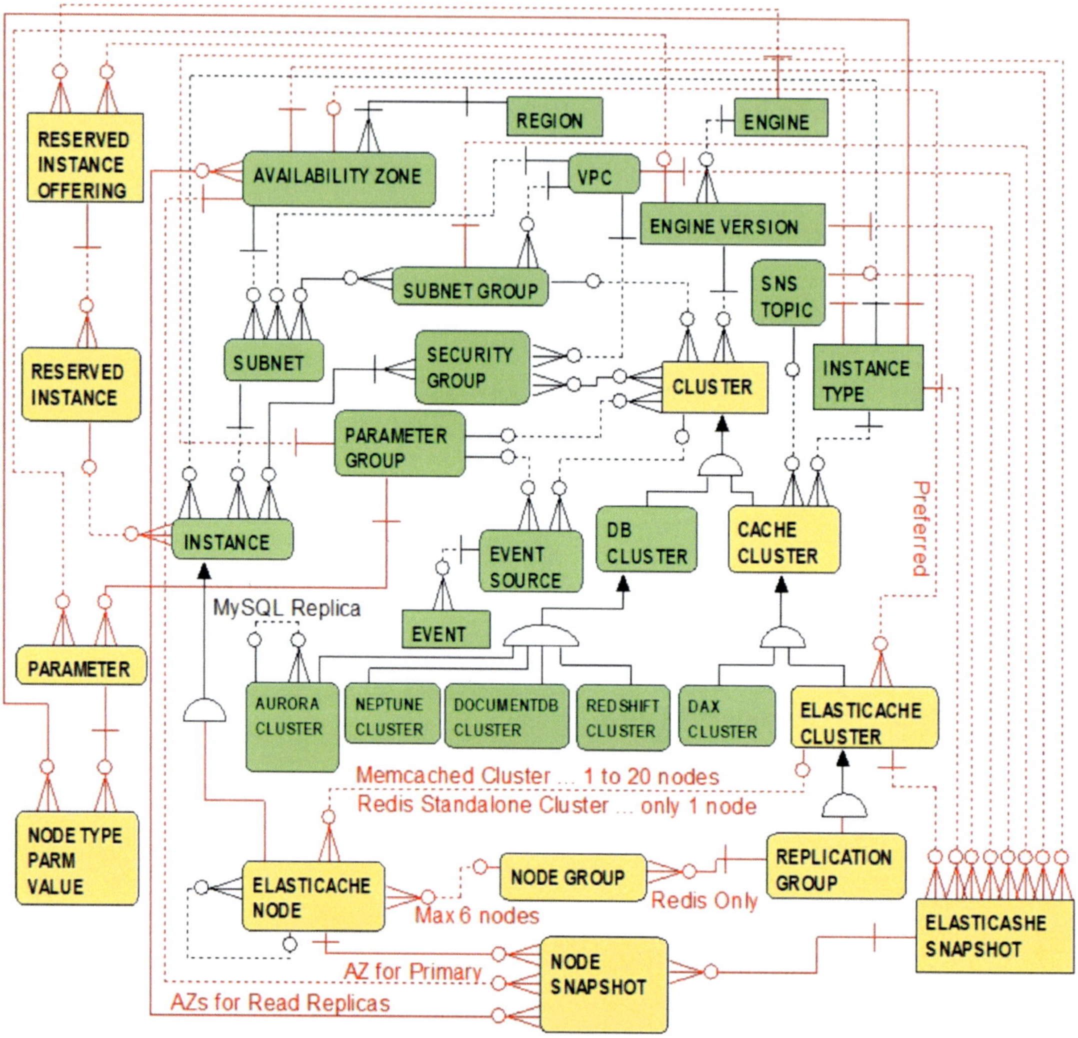

Figure 8.5 ElastiCache Basic Concepts

Analogous to DAX, Amazon ElastiCache is another managed caching service that facilitates the implementation of scalable clusters of cache nodes -- offering the prospect for improving latency by responding to queries with in-memory data and minimizing access to HDDs or SSDs. ElastiCache supports two NoSQL open-source caching strategies: Memcached and Redis. Both implement a key-value store, partitioning across multiple nodes, high availability among multiple availability zones, and capacity scaling.

Redis also flaunts data encryption, read replicas, automatic failover, and backups for recovery of in-memory data. ElastiCache is similar to the DAX but different relating to interfacing with a back-end database. Using a *lazy loading* strategy, an app requests data from an ElastiCache cluster (Redis or Memcached API commands). A *cache hit* returns in-memory data to the app. A *cache miss* (including *stale* data with TTL expiration) results in the app following up with an additional request for directly accessing the database -- and appropriate API for refreshing cache data. An app using ElastiCache might also issue separate DML statements for accessing a possible back-end database. ElastiCache interfaces only with the application and is oblivious of any back-end database. Using a *write through* caching strategy, the app inserts/updates data in its database and also updates the cache (i.e., cache is always current).

Redis: Redis focuses on manipulating data within a cache server's RAM and diminishes the need for a back-end database. Redis also allows for storing in-memory cache data on disk (e.g., via back up snapshots) for persisting in-memory data. Redis CRUD operations are performed in-memory. Redis in-memory changes can also be written to log files. Snapshots and log files serve as the basis for recovering in-memory data for restarting servers or automatic failover to a read replica.

Redis deals with higher level *abstract data types* (transcending RDBMS table/column concepts). Redis (beyond a conventional key-value store) is a data structures server. Redis provides a set of commands specific to each of its supported data types. Redis abstract data structures include: Strings, Lists, Sets, Sorted Sets, Hashes, and Pub/Sub. Redis equips each data type with a distinct set of CRUD commands (e.g., for Strings, CRUD commands include SET, GET, SETRANGE, GETRANGE, MSET, MGET, INCRBY, DECRBY, and APPEND).

Memcached: Memcached -- not a data structure server -- provides a conventional key-value store (i.e., no abstract data types … strings are the only supported data type). Memcached enables a large in-memory hash table (i.e., a key string value is mapped to another string value). Values moved to Memcached strings must be encoded (e.g., pre-serialized using Base64). Like Redis, Memcached APIs for manipulating string values are supported by many programming languages. Examples of Memcached string related commands include: Set Data, Add Data, Replace Data, Append Data, Prepend Data, Get Data, Delete Key, Incr, and Decr.

ElastiCache Clusters: An ElastiCache cluster is *a kind of* cache cluster which in turn is *a kind of* cluster. Because an ElastiCache cluster is *a kind of* cache cluster, an ElastiCache cluster inherits a single instance type (i.e., all nodes in an ElastiCache cluster have the same node type) and an SNS topic for communicating event notifications. An ElastiCache cluster indirectly inherits (because a cache cluster is *a kind of* cluster) many security groups, a subnet group, a parameter group, an engine version, and is a source for event notifications. Unlike a DAX cluster, an ElastiCache cluster can be associated with a single *preferred availability zone* in which cluster nodes are desired to be located (i.e., if not in multiple availability zones).

ElastiCache Nodes: An ElastiCache cluster controls a group of one or more ElastiCache node instances but the cardinality rules are subtly different for Memcached clusters (max of 20 nodes) vs. Redis clusters. If two or more nodes for Memcached, data is partitioned among multiple Memcached nodes (i.e., each node contains a subset of key-value pairs -- the concepts of primary and read replica nodes is moot). A hashing algorithm chooses the target node for accessing a key-value pair. For Redis, a cluster of only one node is a *Redis standalone cluster*.

A Redis cluster of two or more nodes is a *Replication Group* (i.e., *a kind of* ElastiCache cluster) which enables replicating in-memory data and partitioning data across multiple Redis nodes. A replication group clusters Redis nodes into one or more *node groups* (aka *shards*). Each node group consists of a max of 6 nodes -- re-introducing the concepts of a *primary node* and up to 5 *read replica nodes*. For a *Redis (cluster mode disabled) cluster*, a single node group supports replication but disables data partitioning (i.e., partitioning requires at least 2 node groups). A *Redis (cluster mode enabled) cluster* partitions a max of 15 node groups. A node group with a primary node and at least one replica node enables automatic failover in the event of primary node failure.

ElastiCache Backups: An ElastiCache cluster can be backed up via *snapshots*. ElastiCache snapshots apply only to Redis clusters. A snapshot can backup many ElastiCache nodes and preserves *node group configuration* info (e.g., availability zones and *keyspace* for partitioning) for each node included in the snapshot. Only one node from each node group requires backup. For a Redis (cluster mode enabled) cluster, the primary node of each node group is backed up. For a Redis (cluster mode disabled) cluster, any node within the single node group can be selected. This allows backing up a read replica instead of the primary node, enabling the primary node to respond to requests unperturbed by the backup process.

Chapter 9 – Amazon Quantum Ledger Database

Amazon QLDB Introduction

You, the relentless reader and indefatigable data professional (in the final lap of this excursion into cloud database services), are now challenged with fathoming yet another noteworthy and innovative database paradigm -- compliments of the Amazon Quantum Ledger Database service (Amazon QLDB). Amazon QLDB, an additional managed AWS database service, is remarkably different from previously described AWS database services.

QLDB's appeal and purpose is focused on applications requiring a *ledger* database.[65] Briefly, a *ledger* is used in financial accounting for recording business transactions in terms of debits and credits to accounts. Ledgers are also frequently paired with the concept of *journals* where initial transaction details are sequentially recorded (i.e., as accounting *entries*) prior to being posted to ledger accounts.[66]

For QLDB, a ledger database is a data repository for System of Record (SOR) applications requiring complete, accurate, sequential, and unambiguous transaction history.[67] Such repositories are trusted to be the authoritative, single-version-of-truth, and secure source of data for an enterprise's mission critical and proprietary information. Obvious examples of SOR apps crucially requiring trustworthy history include …

- **eCommerce** Online Order Tracking and Fulfilment: Requiring detailed process history from order entry, item picking, packing, and shipping logistics through final delivery.
- **Banking**: Deposits, withdrawals, and transaction history of bank accounts.
- **Product Recall Tracking**: Tracking the bill-of-material (as-built), change history of component parts, and maintenance of complex manufactured product assemblies (e.g., aircraft engines, automobiles, etc.).

The challenge of maintaining verifiable records for proprietary and mission critical data is far from trivial. Techniques for implementing change history for conventional database paradigms include extra effort required for …

- Custom software development (e.g., for triggers, stored procedures, and audit reports in relational systems).
- Database enhancements for maintaining change history (e.g., implementation of partitioned tables, separate history tables, log files, and audit trails).
- Database administration effort for assuring high availability and disaster recovery (e.g., automatic replication of data to backup repositories).

Despite the above extra effort, C-level executives continue to be restless and nervous about data confidentiality, integrity, and availability. CxO sleepless nights are provoked by cyber-security worries surrounding fraud, espionage, sabotage -- from both external and internal threats. Privileged users (e.g., system administrators, super users, users with root access, etc.) can surreptitiously and fraudulently compromised data integrity, access intellectual property, or modify proprietary software. So, additional effort to allay such concerns can result in pricey policies, processes, and people (e.g., for implementing separation of duties -- SOD) to help prevent fraud, error, data loss, or data exfiltration.[68]

Examples of QLDB features mitigating such concerns and significantly distinguishing QLDB from previously described database services are summarized in subsequent subsections.

SQL-based Access to Document Data

QLDB blends familiar concepts from both relational and document database approaches for maintaining ledger information. Specifically, the foundation of a QLDB ledger is defined in terms of one or more *tables* -- indeed reminiscent of RDBMS tables but also similar to DynamoDB's reuse of the table concept for its TABLE entity (as previously described in Chapter 7). However, unlike RDBMS and DynamoDB terminology (i.e., where tables respectively consist of one or more rows or items), each ledger table in QLDB consists of one or more documents reminiscent of document database engines such as DocumentDB (as previously described in Chapter 5). One might describe QLDB as a quasi-relational-document approach for ledger databases.

QLDB documents are similar to DocumentDB documents. Both are schemaless and support an open content JSON-like file format. Furthermore, QLDB's document format, known as Amazon Ion, is actually a superset of the JSON open-standard format and is notably more robust by supporting data types beyond those of DocumentDB.[69] Examples of additional data types offered by Amazon Ion include *blobs*, *clobs* (i.e., for **b**inary **l**arge **ob**jects and **c**haracter **l**arge **ob**jects) and new data types for more precisely specifying numbers (i.e., as integers, decimals, and floating point numbers).

Respectful of SQL's global familiarity and reminiscent of SPARQL's SQL-like approach for accessing Neptune graph data (as previously described in Chapter 4), the QLDB data manipulation language for creating, reading, updating, and deleting documents in ledger tables is also comparable to SQL (i.e., SQL basic DML commands: INSERT, SELECT, UPDATE, DELETE). In particular, the QLDB DML is enabled by PartiQL, an open source query language, developed by Amazon. PartiQL was expressly built for flexibly accessing diverse data formats including relational tables, semi-structured document data, and complex nested documents. Furthermore, PartiQL supports extensions to SQL to streamline access to QLDB documents. PartiQL tutorials and detailed specifications are readily available online.[70]

Especially noteworthy is PartiQL's extended SQL syntax which includes a *dot notation* and *aliasing* of nested data. This enables direct, intuitive access to multi-level, nested data embedded within documents. A simple familiar example of a nested document is a grouping of *PO Items* nested within a *Purchase Order*. The following illustrates a PartiQL INSERT statement for inserting a simple Purchase Order document in ION format …

```
INSERT INTO PurchaseOrder
    {
    'POId' : 'PO123456789',
    'CustomerId' : 'Any Random Customer',
    'OrderDate' : `2019-12-25T`,

    'POItems' :
        [
        { 'ItemId' : 'Random Widget A' , 'Qty' : 1,  'UnitPrice': 1.75},
        { 'ItemId' : 'Random Widget B' , 'Qty' : 2,  'UnitPrice': 2.75},
        { 'ItemId' : 'Random Widget C' , 'Qty' : 3,  'UnitPrice': 3.75}
        ]
    }
```

The indented list of 'POItems' implies a one-to-many relationship -- that the Purchase Order requests one or more purchase order items. The traditional relational approach creates separate tables for PurchaseOrder and POItem with referential integrity enabling joins between the two tables. PartiQL supports table joins but also (via its dot notation and aliasing) promotes direct linkage to nested data. This concisely enables selection, filtering, and aggregating nested data within complex documents. This approach provides an option for fewer tables and obviates convoluted join conditions especially for documents embedding multiple levels of nested data. The following PartiQL SELECT statement is a simple example illustrating access to nested data with "poi" as the alias for the nested "POItems" in the above document …

```
SELECT po.POId, po.OrderDate, poi.ItemId, poi.Qty
FROM PurchaseOrder AS po, @po.POItems AS poi
WHERE po.CustomerId = 'Any Random Customer'
```

In essence, PartiQL's aliasing and dot notation elevates the stature of nested data by more succinctly, intuitively, and efficiently traversing and flattening parent/child associations within documents. This is reminiscent of Gremlin's graph query language paradigm (as described in Chapter 4) for elevating edge relationships to higher importance for simplifying access to highly connected data.

Built-in History and Change Data Capture

QLDB obviates custom software development for maintaining transaction history by automatically capturing all application changes to QLDB ledger documents. Software development and DBA effort for designing and implementing traditionally used "bread-and-butter" techniques (e.g., triggers, stored procedures, partitioned tables, and history tables in relational engines) as mechanisms for maintaining transaction history is unnecessary -- is history -- in QLDB.

QLDB documents are also referred to as *document revisions* since a complete history of all revisions to documents is maintained for ledger tables. A QLDB ledger is basically a *journal* of transactions maintained within an *append-only* data repository. A journal maintains a complete, chronological record of all ledger modifications.

When inserting a new document into a ledger table (via a PartiQL INSERT statement, as illustrated above), QLDB stores the document in the journal as the initial document revision with an automatically generated unique identifier (i.e., a document ID) and a version number of 0. Each document revision in the journal is uniquely identified by combining the document ID and version number (i.e., a composite primary key).

When updating one or more fields of an existing document (via a PartiQL UPDATE statement), QLDB appends a new document revision (i.e., an after-image reflecting the changes) into the journal. The new document revision inherits the same document ID of the previous revision but the version number is incremented by 1. Since the journal is an *append only* repository, the previous document revision is not physically altered. An UPDATE statement does not actually modify the existing revision previously situated in the journal, but instead results in committing a new document revision for reflecting the change. The new document revision with a later version number supersedes (without replacing) the previous version and becomes current. Essentially, a journal maintains a log of transactions each of which committed one or more document revisions to ledger tables. Deleting an existing document (via a PartiQL DELETE statement) also results in appending a new document revision, reflecting a logical deletion, into the journal.

Note that QLDB, borrowing again from relational vocabulary, is also endowed with *views* for distinguishing a table's current document data from historical document revisions. The *user* view is the default view for accessing current application data (i.e., the latest, non-deleted document revisions in the journal). Another view, the *committed* view, is similarly used for accessing current application data but also projects automatically generated metadata (e.g., document ID, version, a transaction ID, transaction date/time, etc.). PartiQL also equips its SELECT statement syntax with a *history function* for accessing the history of prior document revisions including both application data and system generated metadata.

Immutability

Clearly from the previous narrative, a PartiQL UPDATE statement for changing a document does not directly affect the current document revision but instead results in constructing a new document revision. A change to the current document revision renders it history in deference to (and superseded by) the changes reflected in the new (now current) revision. All PartiQL INSERTs, UPDATEs and DELETEs are indisputably journalized -- only appended to the journal -- and thus they do not allow for prior history be modified or physically removed.

The automatic, complete and accurate recording of all change history (data and metadata) leads one to surmise that a journal's data integrity is virtually unchallengeable -- meaning that document revisions (both current and history) cannot be intentionally or accidentally modified after being committed to the journal. But how does QLDB do this?

The undergirding conceptual detail for logging transactions (i.e., the data structure and content comprising the journal) exploits concepts and terminology resonating with *blockchain* technology -- yet another database paradigm but radically different from previously described database services. Blockchain technology deals with ledgers that are implemented across a decentralized, distributed network of cooperating servers (i.e., in a peer-to-peer network). The peer servers collaborate and collectively agree (e.g., using a *consensus algorithm*) without relying on any central authority. You, as a proud and esteemed IT professional (unless having lived totally under an RDBMS rock, or like an ostrich with your head stuck entirely in RDBMS sand), are likely well aware of cryptocurrency controversies and implementations of blockchain technology (e.g., Bitcoin and open source blockchain frameworks such as Ethereum and Hyperledger Fabric).[71]

Adopting similar vocabulary, a QLDB journal consists of a *chain of blocks*. Each ledger transaction results in persisting one or more blocks of data, which are serially chained (i.e., linked) and appended to the journal. Application data and system-generated metadata are stored in the chain of blocks. Specifically, each transaction results in committing one or more blocks and each block consists of one or more data *entries*. Document revisions resulting from PartiQL DML statements along with metadata -- including the actual statements that committed document revisions -- are stored as block entries (More about this later ... Describing the detailed structure of each block is deferred to a subsequent section of this chapter). One might pontificate, therefore, that QLDB by blending relational, document and blockchain concepts uniquely represents a quasi relational-document-blockchain database approach.

A journal's transaction log consisting of these chained blocks is intuitively immutable given that QLDB automatically logs history, is innately incapable of modifying or deleting history, and reflects changes to current documents by displacing them with new revisions. But, in addition, QLDB provides a mathematical process for verifying the integrity of document revisions. How does QLDB purport to do this? Read on ...

Verifiability

Recall for ElastiCache replication groups (as previously described in Chapter 8) that a hashing algorithm (i.e., the CRC16 hashing algorithm) facilitates access to cached data partitioned across multiple cache nodes. Hashing techniques are also used by QLDB but for a different purpose -- as the basis for verifying that transaction data has not been transformed, tampered or tarnished in any way since initially committed to the journal.

As previously noted, each document revision maintains both application data and document metadata. In addition to the examples of previously noted document metadata (i.e., document ID, a version number, a transaction ID, etc.), each block also maintains an assortment of SHA-256 (**S**ecure **H**ash **A**lgorithm) hash codes.[72] SHA-256 is a cryptographic hash function -- a mathematical algorithm -- that accepts any given data as input and maps it to a 256 bit hash code (aka *digest*, *signature*, *fingerprint*, *checksum*).[73] In essence, the hash code represents a condensed summarization of the input data. SHA-256 hash codes are the key ingredient for verifying data integrity because of the high likelihood that any changes to the input data will result in a different hash value.[74]

Each block in the journal is equipped with multiple hash codes. Examples of block attributes whose values are SHA-256 hash codes include *block hash* (i.e., the hash value for all data within a block, aka *blockHash*) and *previous block hash* (i.e., the hash value for the previous block in the chain, aka *previousBlockHash*). The *block hash* becomes the *previous block hash* for the next block in the chain and is, therefore, also ingested within the next block's *block hash*. The *previous block hash* is thus inextricably intermeshed with the *block hash*.

Each document revision stored within a block is also summarized with a SHA-256 hash value. QLDB provides a verification process to confirm data integrity for any designated document revision within a block. The QLDB verification process is briefly summarized as follows:

1. **Request a Digest**: A digest is a file (stored in Amazon ION format) and is requested via the Amazon QLDB console or QLDB API. A digest file includes yet another SHA-256 hash, the *digest hash*. A digest hash condenses a ledger's full change history up to and including the journal's last saved block (aka the *tip* of the journal) at the time the digest is requested. A digest file is a key component for verifying a document's data integrity. Any document committed to any block within the digest's scope can be verified.

   A digest file applies the concept of a Merkle Tree -- a concept also shared with blockchain technology.[75] The digest's SHA-256 hash is based on a binary tree structure of hash codes starting with the bottom nodes of the tree which are the SHA-256 hash codes for journal blocks. Block hashes are sequentially paired (e.g., 1st block hash paired with 2nd block hash ... 3rd block hash paired with 4th block hash, etc.) and concatenated. The result is then rehashed to create a new hash node (aka a *proof hash*) at a higher level in the tree structure. Proof hashes at the higher level of the tree are next paired, concatenated and rehashed to higher levels (consisting of new *proof hashes*) as necessary. Figure 9.1 proposes a simple example.

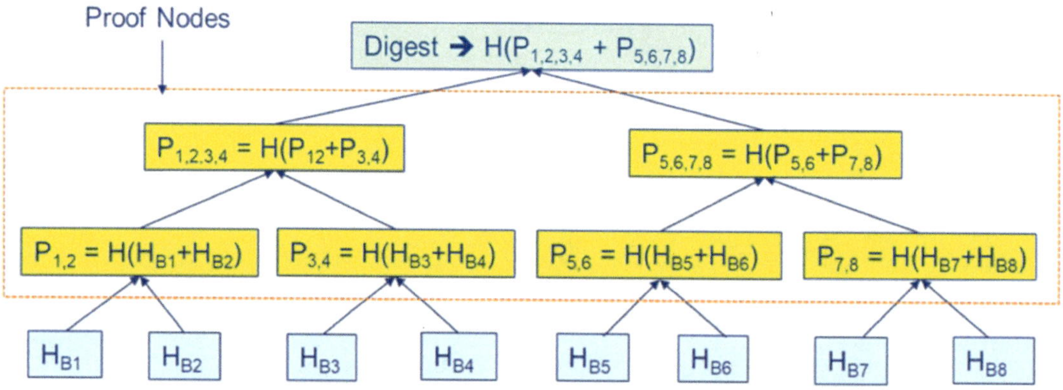

Figure 9.1 Digest Binary Tree ... Simple Example

In Figure 9.1, lowest level nodes (H_{B1}, H_{B2}, etc.) represent block hashes. Next higher level nodes ($P_{1,2}$, $P_{3,4}$, etc.) represent proof hashes from lower level block hashes that have been paired, concatenated, and rehashed. These proof hash nodes are then similarly processed resulting in the next higher level of proof hash nodes ($P_{1,2,3,4}$ and $P_{5,6,7,8}$), which in turn are likewise rehashed. The end result is the digest hash code at the top of the tree.

The digest file also contains a *digest tip address* which identifies the location of the last block consumed in the digest. The saved digest file can then be used for verifying data integrity of any document stored in a block up to an including the last block identified in the digest tip address.

2. **Identify Document To Be Verified:** Metadata for identifying the specific document to be verified must be retrieved from the ledger. As previously noted, a document's system-generated metadata can be accessed via a SELECT statement using the *committed* view. The retrieved metadata includes the unique document id (*metadata.id*) and a block address *(blockaddress).* The block address includes a sequence number *(blockaddress.sequenceNo)* for helping identify the document's block location in the journal.

3. **Start Verification:** The QLDB console provides a *verify document* form enabling a user to specify the document to be verified by entering the ledger name, document id and block address. Form options are also provided for specifying the digest to be used for verification by either selecting an existing digest file or manually entering the digest hash and digest tip address from a saved digest file. The user can then choose to initiate the verification process.

4. **Verification Proof:** QLDB recycles another concept familiar to blockchain technology and cryptography -- Merkle Audit Proofs -- as the foundation for verifying a document's integrity.[76] A Merkle audit proof essentially provides a short-cut method for checking integrity without the need for sequentially traversing the chain of blocks and recalculating hash codes for possibly much of a ledger's entire document history. Instead, a much smaller number of the digest's tree nodes are needed. Specifically, only hashes from a subset of the nodes in the binary tree for the digest (specified in the previous step) participate in the verification process. This subset of hashes is referred to as the *audit path* from the block of the document to be audited (leaf node) to the digest (i.e., the root hash).

   After the user initiates the verification process, QLDB provides what is referred to as a *proof* -- an ordered list of node hashes (from the digest's node tree) for computing (i.e., rehashing) other hashes (for missing nodes not included in the ordered list) at higher levels in the tree. A recalculated hash is then combined with that of another given proof node hash (i.e., the next proof node hash in the ordered list) yielding another recalculated hash, and so on. This tree climbing process ultimately results in recalculating the digest hash. The recalculated digest hash should be equal to the originally calculated digest hash as saved in the digest file -- thus confirming the document's integrity.

   Figure 9.2 modifies the colors in Figure 9.1 highlighting a simple audit path for verifying a document.

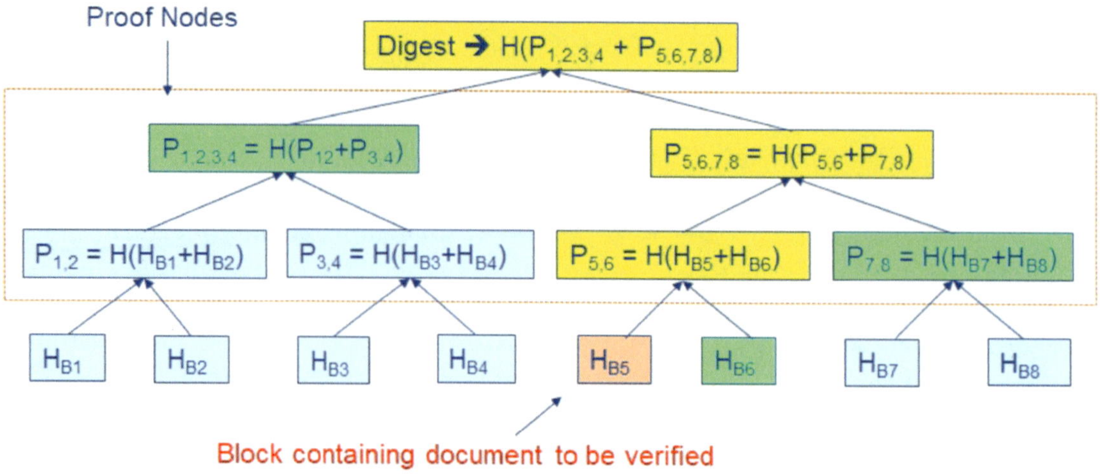

Figure 9.2 Audit Path Example

In Figure 9.2, suppose that node H_{B5} is the block that contains the document whose integrity is to be verified. The ordered list of hashes that QLDB provides (shown in green) are H_{B6}, $P_{7,8}$, $P_{1,2,3,4}$ respectively required to calculate higher level hashes (shown in yellow) $P_{5,6}$, $P_{5,6,7,8}$ and finally the digest at the top of the tree.

QLDB Basic Entities and Relationships

Figure 9.3 introduces QLDB basic entities and relationships.

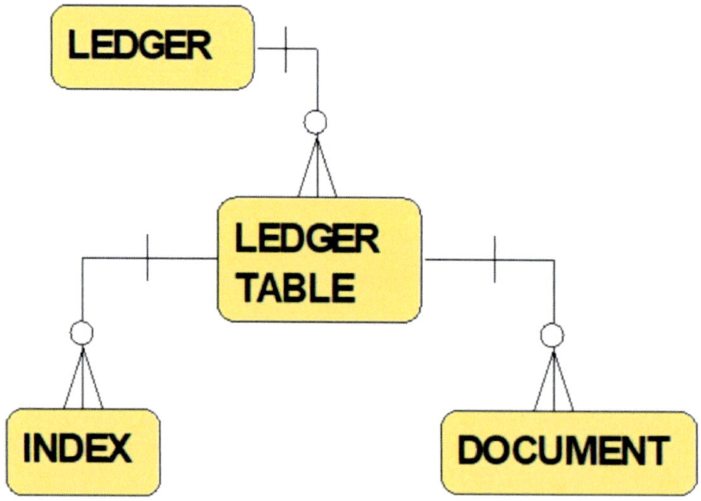

Figure 9.3 QLDB Basic Entities and Relationships

In Figure 9.3, the LEDGER entity reflects the set of all ledger databases controlled by QLDB. A ledger in QLDB is intuitively at the same level as that of a database in a relational engine. QLDB can support many ledgers in the same sense that a relational engine, such as OracleDB, can support many databases. The one-to-many relationship from the LEDGER entity to the LEDGER TABLE entity reflects that each ledger in QLDB is defined in terms of one or more tables, again reminiscent of relational tables -- and also reminiscent of DynamoDB's reuse of the table concept for its TABLE entity (as previously described in Chapter 7).

The one-to-many relationship from the LEDGER TABLE entity to the DOCUMENT entity reflects that each ledger table in QLDB consists of one or more documents -- unlike RDBMS and DynamoDB terminology where tables respectively consist of one or more *rows* or *items*. As previously noted …

- Documents are equivalently described as *document revisions* since a complete history of all revisions to documents are maintained for a table.
- A document ID and version number are automatically generated and combine to uniquely identify each document revision.
- Documents are *schemaless* and obey Amazon Ion format.
- Documents include metadata attributes (in addition to document ID and version number) such as a transaction ID, a timestamp, and SHA-256 hash value.

The one-to-many relationship from the LEDGER TABLE entity to the INDEX entity reflects that each ledger table can be defined with one or more table indexes for more efficiently accessing document data. The INDEX entity is reminiscent of secondary indexes and the SECONDARY INDEX entity defined for DynamoDB (previously described in Chapter 5). Each QLDB table index is a single field in a key-value pair for a document. Tables and indexes can be interactively created via QLDB console and its *Query editor* by respectively exercising PartiQL CREATE TABLE and CREATE INDEX commands.

QLDB Journal Entities and Relationships

Figure 9.4 extends Figure 9.3 by blending it with entities and relationships describing the QLDB journal concepts.

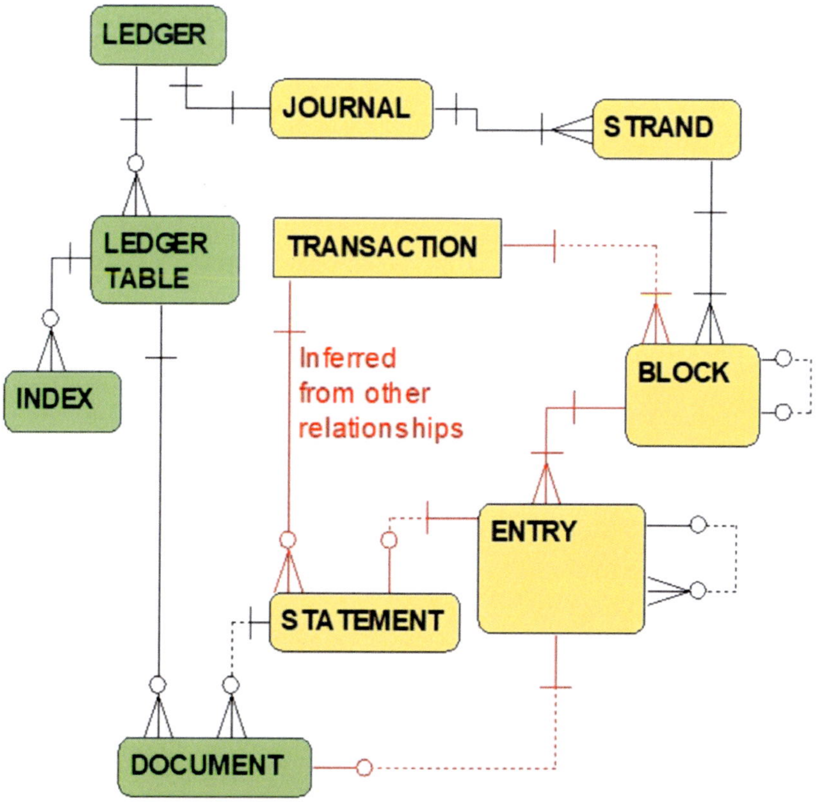

Figure 9.4 QLDB Journal Basic Concepts

In Figure 9.4, the JOURNAL entity reminds us that each QLDB ledger maintains a complete, chronological journal of all ledger transactions. The one-to-one relationship between the LEDGER and JOURNAL entities emphasizes the tight semantic coupling between the concept of a QLDB ledger and a journal. One could argue that they are interdependent -- that a QLDB ledger cannot exist without a journal and that a journal cannot exist unlinked to a ledger (i.e., an *existence dependency*). One could then infer further that both entities should be commingled into a single LEDGER entity -- that there is no need for a separate JOURNAL entity in a conceptual data model. A journal is an indelible component of QLDB ledger but distinct from other ledger components, such as separate QLDB storage supporting table indexes (aka *indexed storage*). Separate LEDGER and JOURNAL entities are proposed in Figure 9.4 in deference to QLDB documentation which appears to regard these as distinct concepts.

The one-to-many relationship between the JOURNAL entity and the STRAND entity reveals that each journal is composed of at least one or more *strands* -- perhaps a method for partitioning journal content into multiple threads. Note that QLDB currently supports only a single journal strand, but one might forecast future support for multiple strands (e.g., a proactive design option for enhancing latency).

The one-to-many relationship between the STRAND entity and the BLOCK entity reveals that each strand is composed of at one or more blocks. The BLOCK entity's one-to-one reflexive relationship conveys the "chain" in the chain-of-blocks concept reminiscent of blockchain technology. Each block can be chained to a prior block in the sequence of blocks. All but the initial block is chained to a prior block and the last block (tip of the journal) indeed has not yet been chained by a subsequent block. Hence, this one-to-one reflexive relationship is optional in both directions, since there exists the possibility of blocks without a predecessor or subsequent block. Therefore, in essence, a journal is simply the historical collection of all the blocks that have been committed in a ledger.

Attributes of the BLOCK entity include the strand id (*strandId*), a block sequence number (*sequenceNo*), transaction id (*transactionId*), a time stamp for the block (*blockTimestamp*), the block hash (*blockHash*), and the previous block's hash (*previousBlockHash*).

The TRANSACTION entity reflects all the transactions executed for committing documents to the journal. The one-to-many relationship between the TRANSACTION entity and the BLOCK entity reveals that each successful transaction generates at least one and possibly many blocks --- but each block must be tied to a single transaction.

So, more precisely, a journal is the historical collection of all the blocks that have been committed by transactions to a ledger. The one-to-many relationship between the TRANSACTION and STATEMENT entities reveals that one or more PartiQL statements were executed for each transaction committed to the journal. Each statement must be tied to a single transaction.

The one-to-many relationship between the BLOCK and ENTRY entities is a reminder that each block, as previously noted, is composed of at least one or more entries, within which transaction details (e.g., document revisions) are stored. Actually, the content of each block is implemented in Amazon ION format. The sample attributes noted above for the BLOCK entity (i.e., *strandId*, *sequenceNo*, *transactionId*, *blockTimestamp*, *blockHash*, and *previousBlockHash*) are included in the block's ION structure as name/value pairs (aka fields). Additional entries are included with nested lower level entries. Examples include:

- A *revisions* entry with a nested list of document revisions (i.e., nested entries) each containing a document's application data and metadata.
- A *transactionInfo* entry with a nested list of PartiQL statements associated with the block's transaction.

Therefore, the ENTRY entity in Figure 9.4 connotes a general term for both high level entries and lower level nested entries (nested data objects) within a block. The ENTRY entity's one-to-many reflexive relationship portrays this hierarchy of entries that can be composed of lower level entry objects.

The one-to-one relationship between the ENTRY entity and the DOCUMENT entity reveals that an entry can map to a table document (i.e., a single document revision), but that a document must be tied to a single entry of a block. In reality, previously noted attributes for the DOCUMENT entity (e.g., document ID, version number, SHA-256 hash value, etc.) are implemented as fields within the document's associated block entry.

Likewise, the one-to-one relationship between the ENTRY entity and the STATEMENT entity reveals that an entry can map to a single statement executed for the transaction, but that each statement must be tied to a single entry of a block. Fields for the statement in the entry includes the text of the statement itself and its start time.

The STATEMENT entity, therefore, is indirectly associated with the TRANSACTION entity via its relationship to the ENTRY entity. The one-to-many relationship between the TRANSACTION and STATEMENT entities is really an *inferred* relationship. It in fact is redundant because it can be inferred from other relationships (i.e. relationships navigating from STATEMENT to ENTRY to BLOCK and finally to TRANSACTION entities). Figure 9.4 highlights that the one-to-many relationship between the TRANSACTION and STATEMENT entities is an inferred relationship.

QLDB Digest Entities and Relationships

Figure 9.5 extends Figure 9.4 with entities and relationships describing QLDB digest concepts.

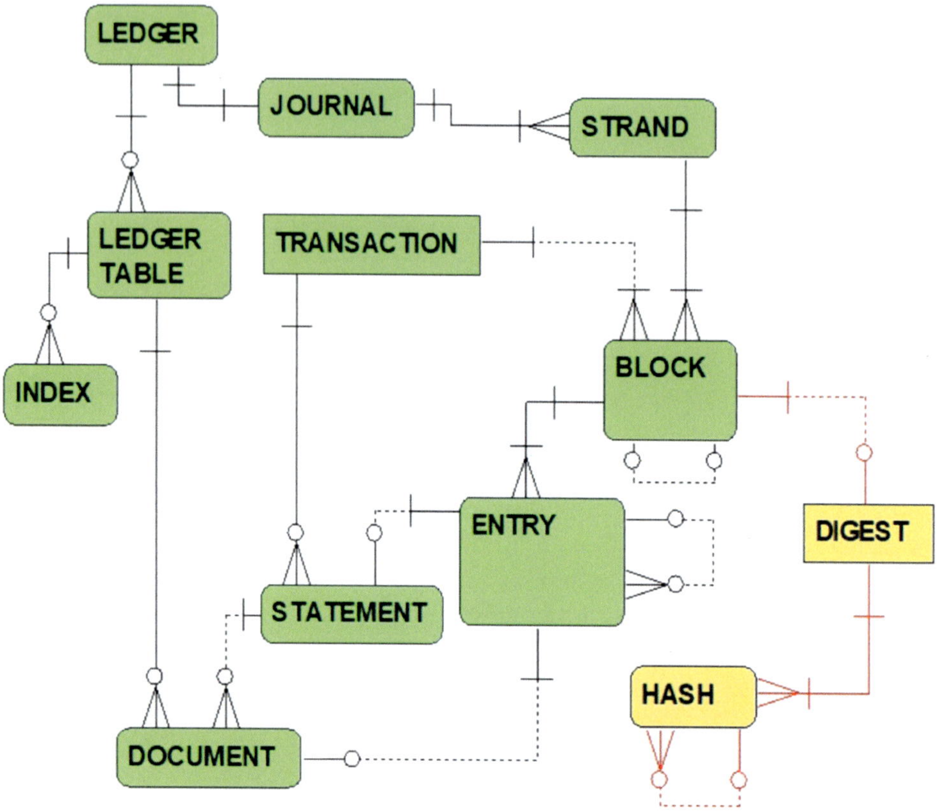

Figure 9.5 QLDB Digest Basic Concepts

In Figure 9.5, the DIGEST entity represents the set of all digest files requested via the Amazon QLDB console or QLDB API. The one-to-one relationship between the BLOCK and DIGEST entities reflects that each digest identifies the last block (i.e., tip of the journal) at the time the digest is requested. The one-to-many relationship between the DIGEST and HASH entities emphasizes that each digest is associated with a binary tree structure of SHA-256 hash codes including lowest level block hashes and higher level proof hashes culminating in the calculation of the digest hash at the tree summit. The one-to-many reflexive relationship between the HASH entity and itself reaffirms the Merkle tree structure among hashes.

Chapter Summary

Figure 9.6 summarizes QLDB concepts addressed in this chapter.

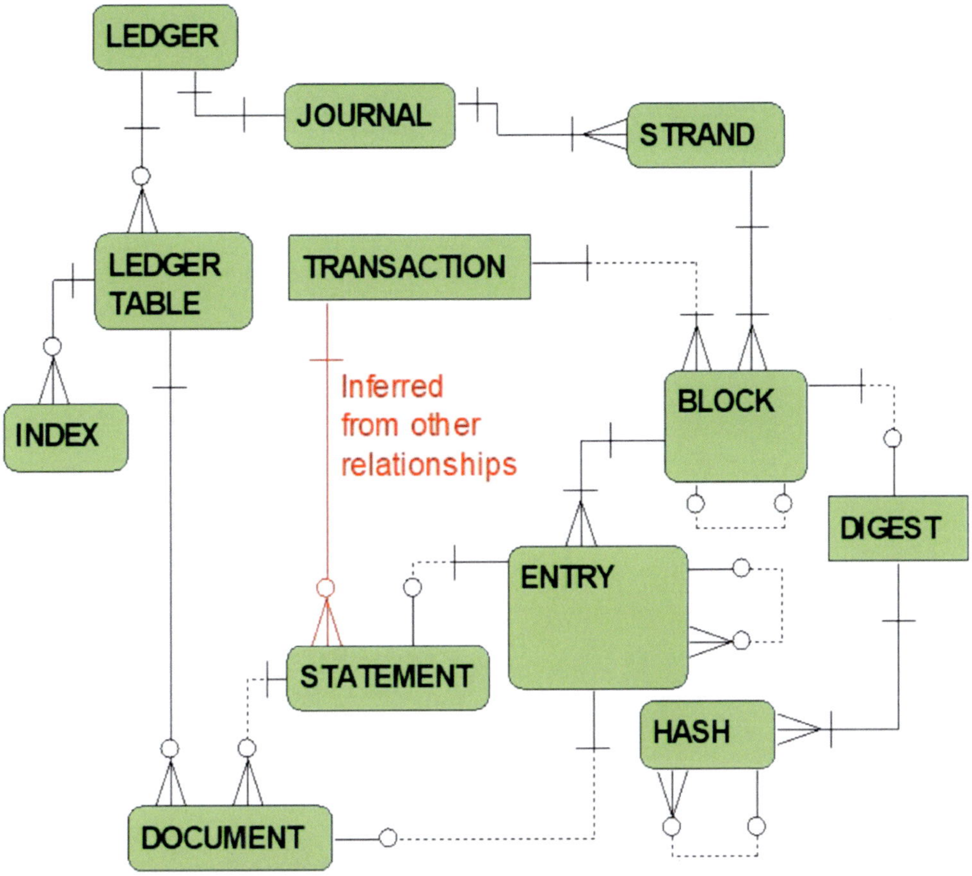

Figure 9.6 QLDB Concepts Summary

Amazon QLDB, a managed AWS database service, is remarkably different from previously described AWS database services. QLDB focuses on apps requiring a *ledger* database. For QLDB, a ledger database is a data repository for System of Record (SOR) apps requiring complete, accurate, transaction history. Examples of SOR apps include eCommerce (online order tracking), Banking (account transaction history), and Product Recall (tracking change history of complex product assemblies). Techniques for implementing change history for conventional database paradigms include extra effort required for custom software development, database enhancements, and database administration. Effort is also required to mitigate internal cyber-security threats to data confidentiality, integrity, and availability (e.g., for implementing separation of duties … SOD).

QLDB features distinguishing it from previously described database services include SQL access to documents, built-in change data capture, immutability, and verifiability.

SQL-based Access to Documents: A QLDB ledger is defined in terms of one or more *tables*. Unlike RDBMS and DynamoDB terminology (i.e., where tables respectively consist of one or more rows or items), each ledger table consists of one or more documents (i.e., similar to DocumentDB).

QLDB documents are *schemaless* and formatted in Amazon Ion, a superset of the JSON open-standard format. Amazon ION extensions to JSON files include *blobs*, *clobs* and data types for specifying integers, decimals, and floating point numbers. The QLDB DML is PartiQL, an open source query language developed by Amazon. Comparable to SQL, PartiQL was expressly built for flexibly accessing a variety of data formats including relational tables, semi-structured document data, and complex nested documents. PartiQL supports an extended SQL syntax. Its extensions include a *dot notation* and *aliasing* of nested data, enabling direct access to nested data embedded within documents (e.g., *PO Items* nested within a *Purchase Order*).

Built-in History and Change Data Capture: QLDB automatically captures changes to ledger documents. Traditional mechanisms for maintaining transaction history are obviated (e.g., triggers, stored procedures, partitioned tables, etc.). QLDB documents are also referred to as *document revisions* since a complete history of all revisions is maintained for ledger tables. A QLDB ledger is basically a *journal* of transactions maintained within an *append-only* data repository.

Each document revision inserted into a ledger table (via a PartiQL INSERT statement) is uniquely identified by combining an automatically generated document ID and version number -- a composite primary key. When an existing document is updated (via a PartiQL UPDATE statement), QLDB appends a new document revision to the journal (same document ID, but the version number is incremented by 1). The previous document revision is not physically altered. A new document revision for reflecting the change supersedes (without replacing) the previous version and becomes current. Deleting an existing document (via a PartiQL DELETE statement) also results in appending a new document revision into the journal.

QLDB supports *views*. The default *user* view is used for accessing current application data, but no metadata. The *committed* view enables access to current info and also includes automatically generated metadata (e.g., document ID, version, a transaction ID, transaction date/time, etc.). PartiQL also provides a *history function* for accessing the history (app data and metadata) of prior document revisions.

Immutability: All PartiQL INSERTs, UPDATEs and DELETEs are journalized such that data can only be appended to the journal -- not allowing prior history to be modified or physically removed. Document revisions (both current and history) cannot be modified after being committed to the journal.

The journal data structure recycles concepts and terminology resonating with *blockchain* technology. A QLDB journal consists of a *chain of blocks*. Each ledger transaction results in persisting one or more blocks of data -- serially chained (i.e., linked) and appended to the journal. Application data and system-generated metadata are stored in the chain of blocks. Each transaction results in committing one or more blocks and each block consists of one or more data *entries*. Document revisions, metadata, and the actual PartiQL statements that committed document revisions are stored as block entries.

A journal's transaction log consisting of these chained blocks is intuitively immutable given that QLDB automatically logs history, does not modify or delete history, and reflects changes by appending new revisions.

Verifiability: QLDB uses hashing techniques for verifying that transaction data has not been transformed since initially committed to the journal. Each block maintains an assortment of SHA-256 (**S**ecure **H**ash **A**lgorithm) hash codes. SHA-256 is a cryptographic hash function that maps any given data input to a 256 bit hash code -- a condensed summarization of the input data. Altering input data results (with high likelihood) in a different hash value -- the key for verifying data integrity. Examples of block attributes whose values are SHA-256 hash codes include *block hash* (i.e. the hash value for all data within a block) and *previous block hash* (i.e., the hash value for the previous block in the chain). The *block hash* becomes the *previous block hash* for the next block in the chain. Each document revision stored within a block is also summarized with a SHA-256 hash value.

QLDB's data integrity verification process is summarized as follows:

1. **Request a Digest**: A digest is a file (in Amazon ION format, requested via the Amazon QLDB console or QLDB API) that includes yet another SHA-256 hash, the *digest hash*. A digest hash condenses a ledger's full change history up to the last block of the journal. Any document committed to any block within the digest's scope can be verified. A digest file uses a binary tree structure (Merkle Tree) to calculate the digest's SHA-256 hash. Hash codes for blocks are paired, concatenated, and hashed resulting in *proof hashes* at a higher level in the tree structure. Proof hashes are similarly combined for higher level proof hashes -- ultimately resulting in the digest hash code at the top of the tree. The digest file also contains a *digest tip address* which identifies the location of the last block consumed in the digest.

2. **Identify Document To Be Verified:** Metadata for identifying the specific document to be verified must be retrieved from the ledger -- the unique document id (*metadata.id*) and a block address *(blockaddress).*

3. **Start Verification:** The QLDB console provides a *verify document* form enabling a user to start the verification process by specifying the document to be verified (ledger name, document id and block address) and either selecting an existing digest file or manually entering the digest hash and digest tip address from a saved digest file.

4. **Verification Proof:** QLDB uses the concept of a Merkle Audit Proof for verifying a document's integrity. Hashes from only a subset of the hash nodes in the digest's binary tree are used. This subset of hashes is referred to as the *audit path* from the block of the document to be audited (leaf node) to the digest (i.e., the root hash). QLDB provides a *proof* -- an ordered list of node hashes (from the digest's node tree) for computing (i.e., rehashing) other hashes (for missing nodes not included in the ordered list) at higher levels in the tree. The process of recalculating hashes climbs up the tree -- resulting in recalculating the digest hash. A recalculated digest hash equal to the originally calculated digest hash as saved in the digest file confirms the document's integrity.

Appendix – Summary of Summary Diagrams

Someday, perhaps very soon, even if you currently are -- or eventually become -- completely confident, competent and conversant with the details of one or more Amazon Database Services, you may be motivated (or induced) to revisit these concepts for any of a variety of reasons. For example, after moving full speed ahead to migrate your existing relational databases to RDS, Aurora, or Redshift, your leadership undoubtedly will next exhort you to blaze new trails beyond your relational comfort zone, such as kicking the tires of graph databases, document databases, serverless databases, ledger databases, caching servers, etc.

Summary ERDs for each of Chapters 1 through 9 are replicated here, as a matter of convenience, to facilitate periodic and relatively rapid refreshment of your database services confidence and comprehension.

In addition, stay tuned to the author's social media for information about viewing and obtaining a physical (e.g., 24 x 36) summary diagram, integrating all Amazon Database Services essential concepts, by melding all summary diagrams of Chapters 1 through 9 within a single ERD. Evidence of the existence of such is offered as follows in Figure A.1, a compressed summary diagram also serving as a link for providing status information of printed-copy availability.

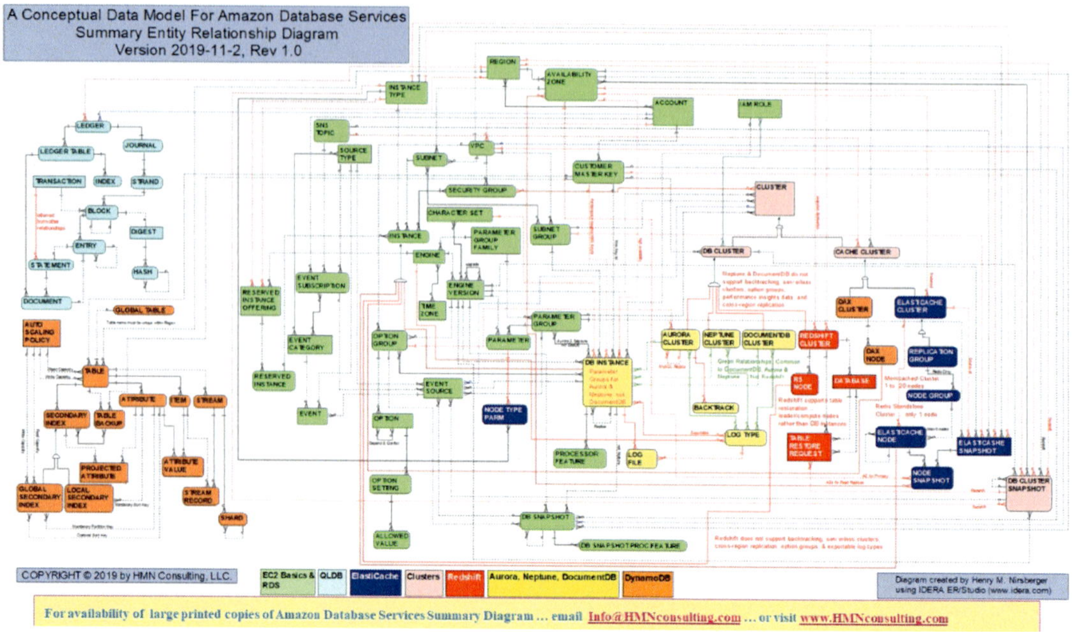

Figure A.1 Conceptual Data Model For Amazon Database Services Summary Diagram[77]
For availability of printed copies of summary diagram ...
email Info@HMNconsulting.com or visit www.HMNconsulting.com

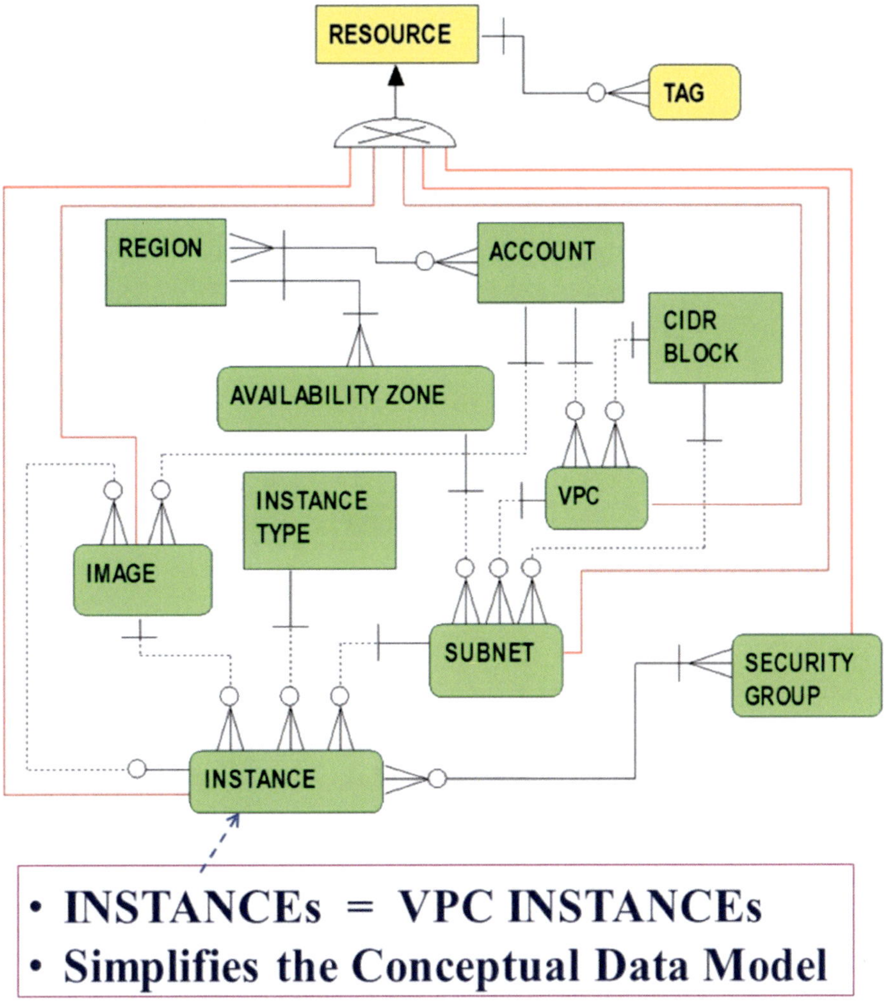

Figure A.2 Amazon EC2 Basics Summary

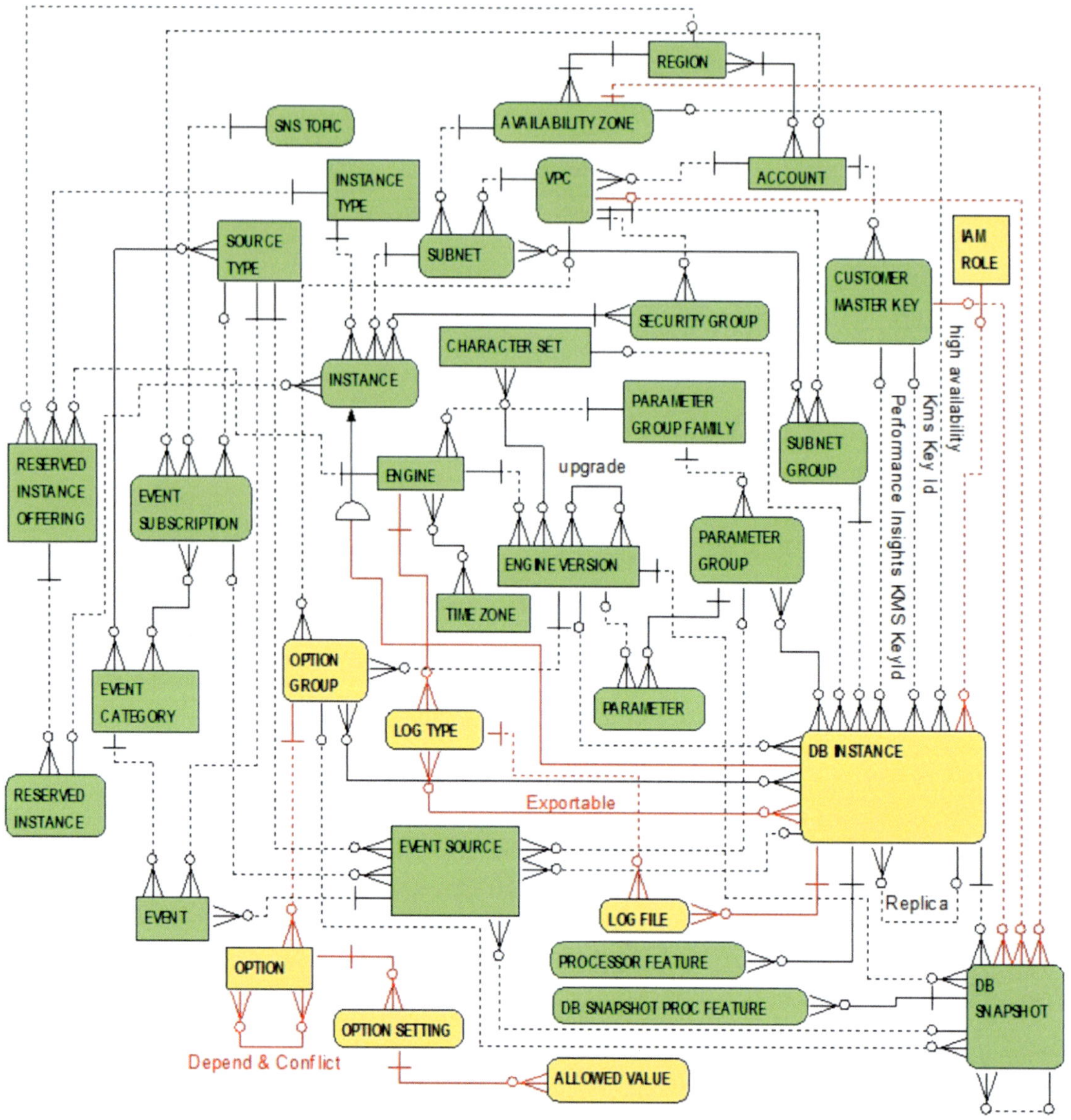

Figure A.3 RDS Summary

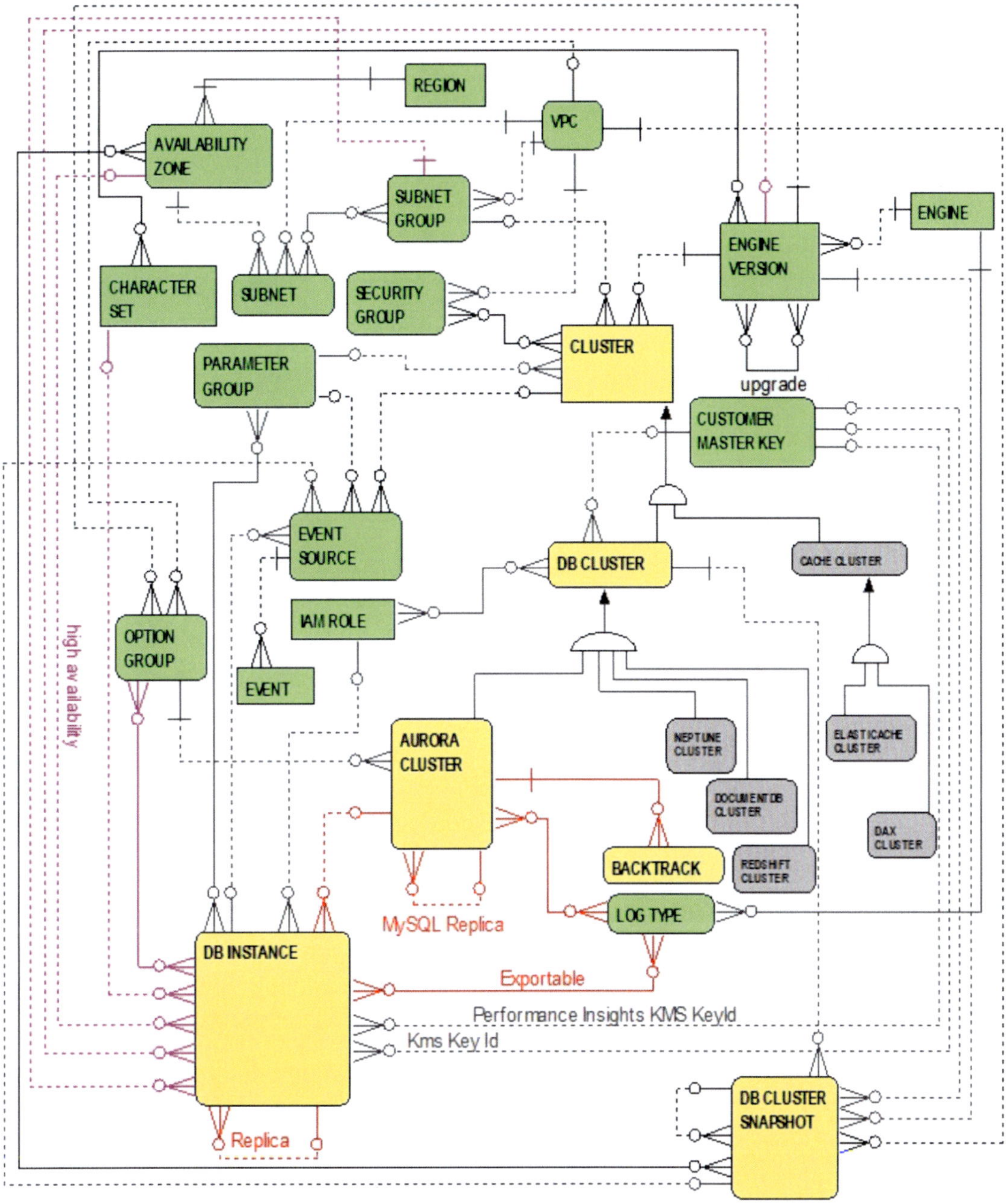

Figure A.4 Aurora Summary

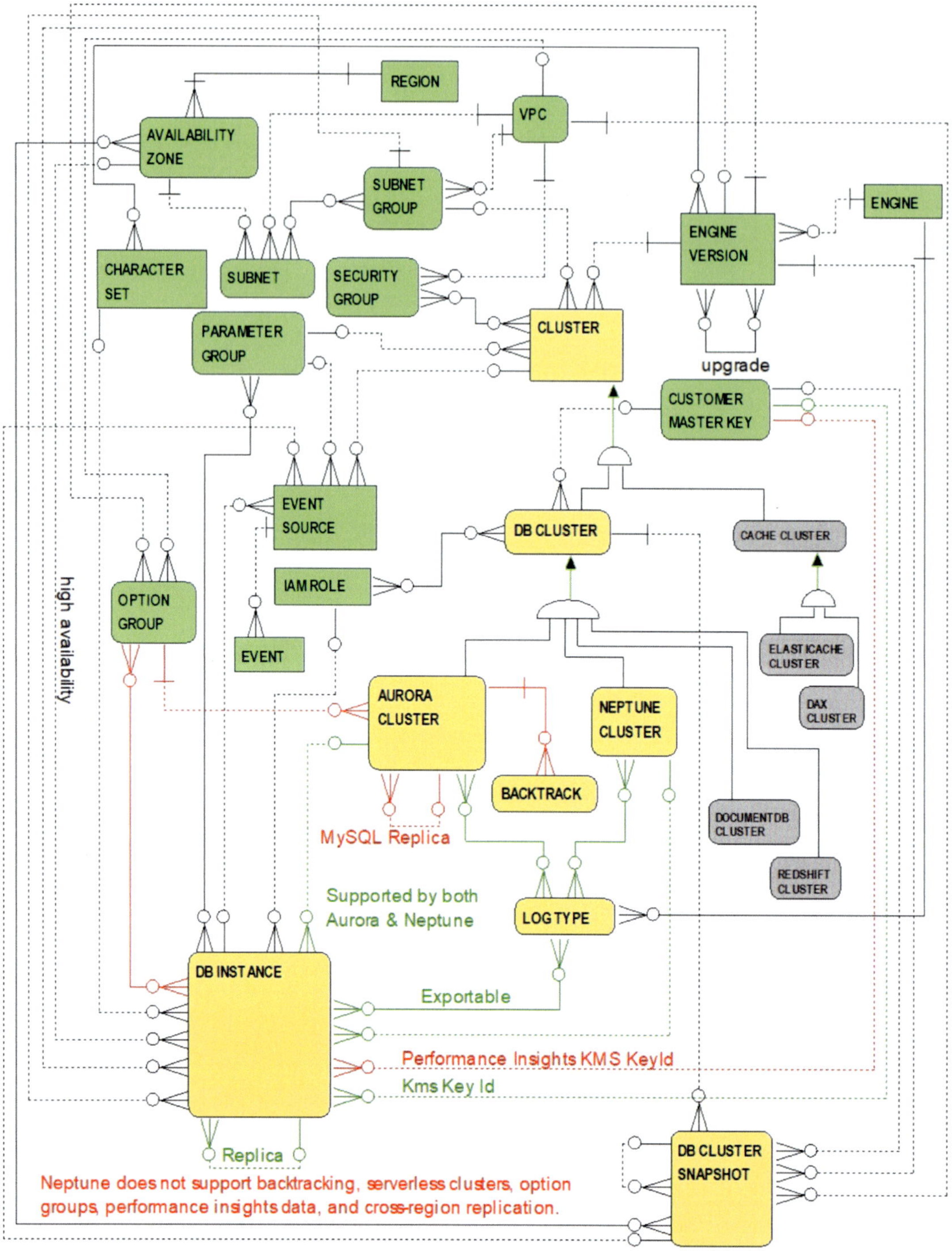

Figure A.5 Neptune Summary with Aurora Similarities and Differences

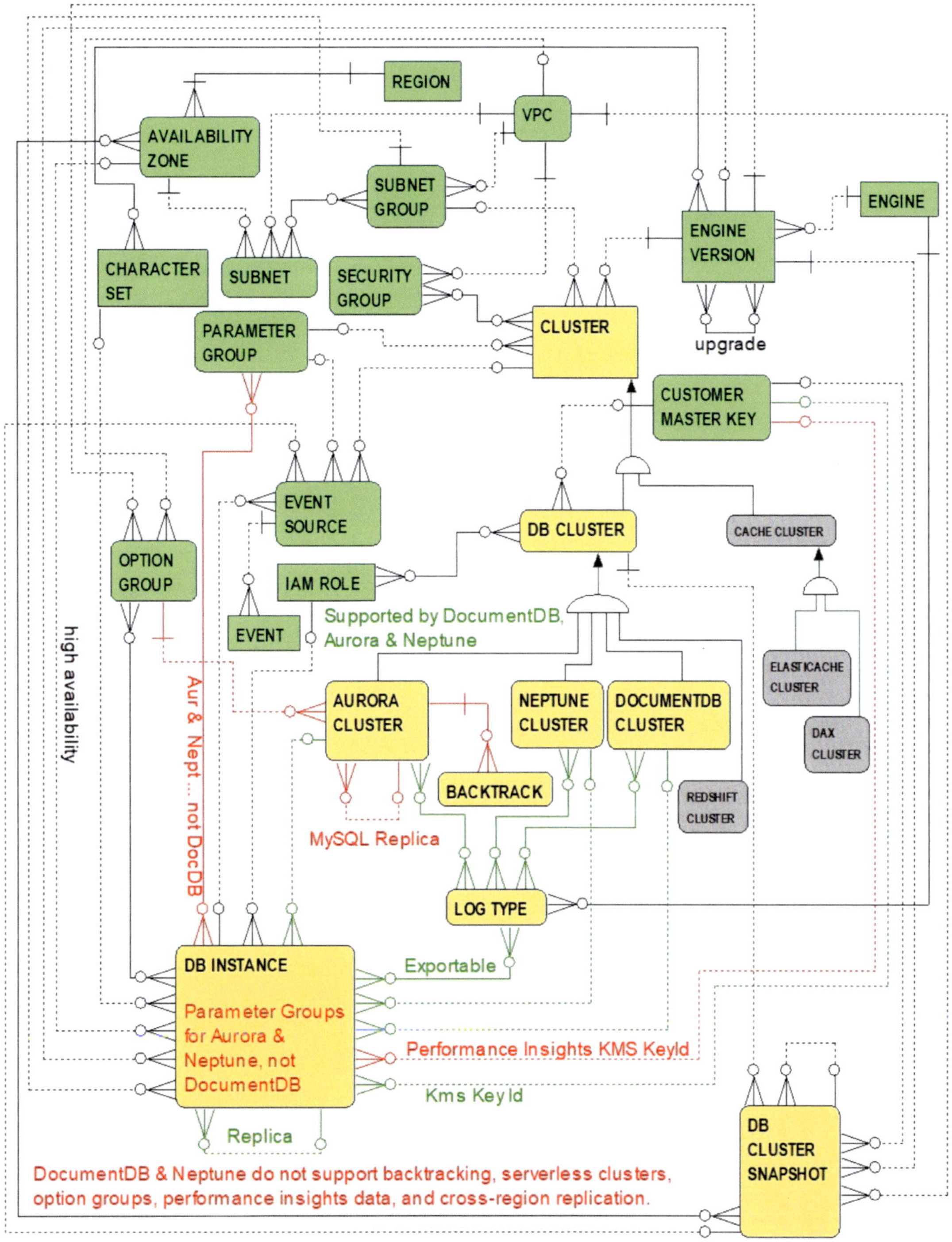

Figure A.6 DocumentDB Summary Compared with Aurora and Neptune

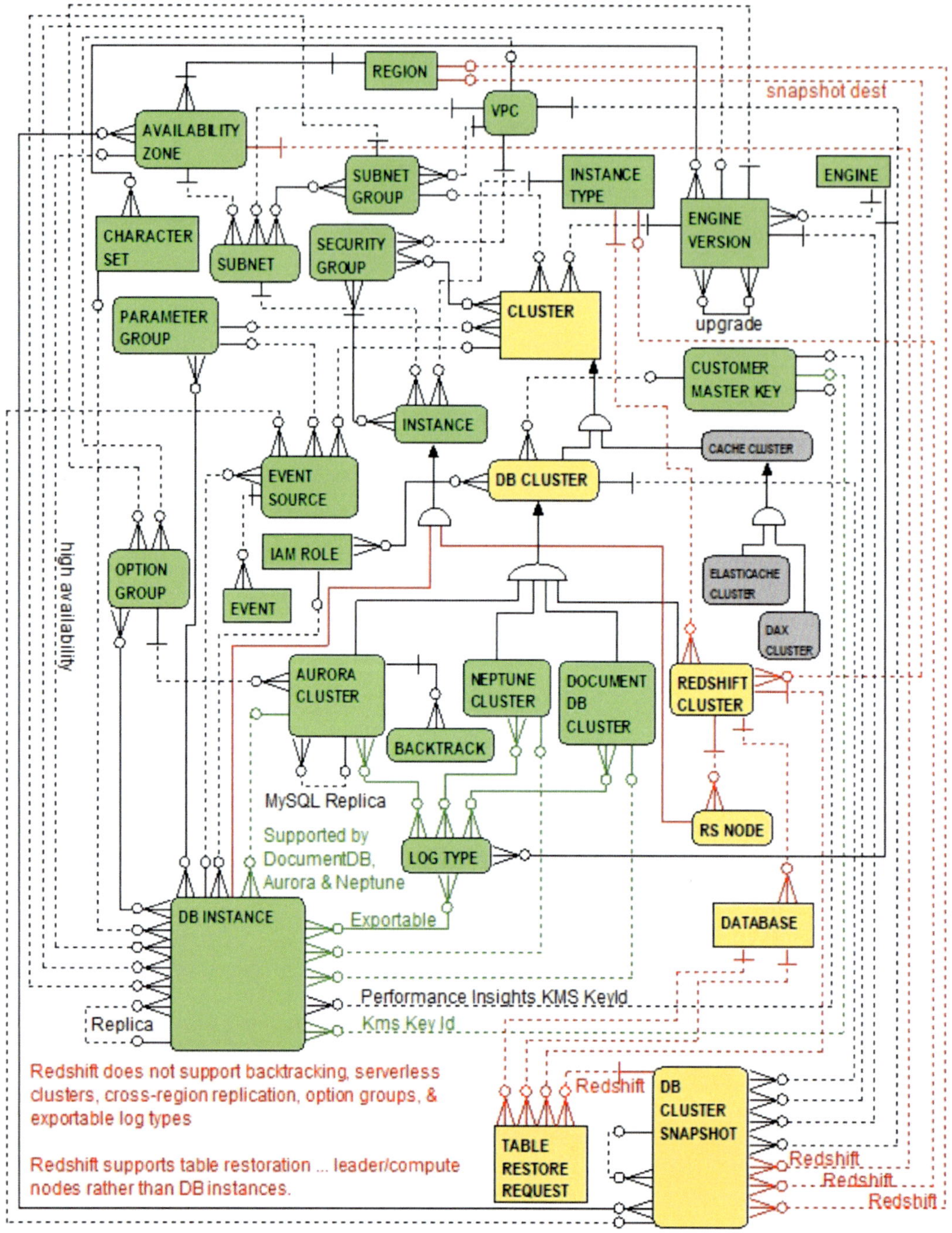

Figure A.7 Redshift Summary Contrasting Aurora, Neptune and DocumentDB

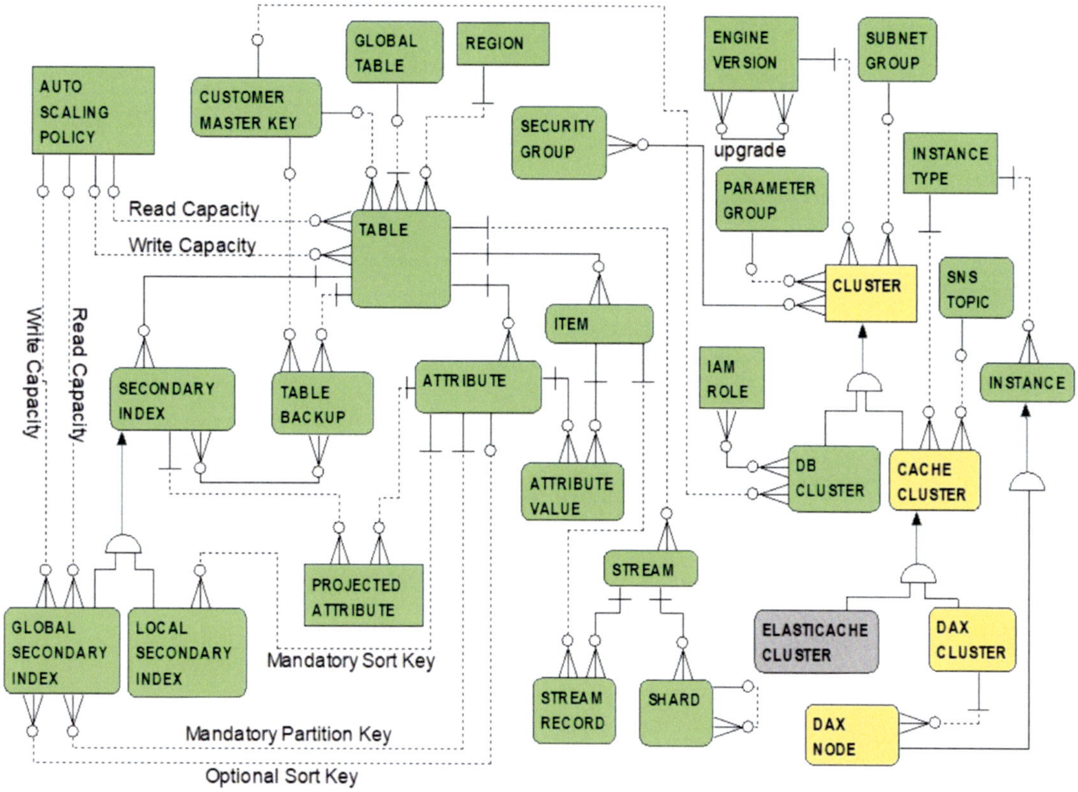

Figure A.8 DynamoDB Entity Relationship Summary

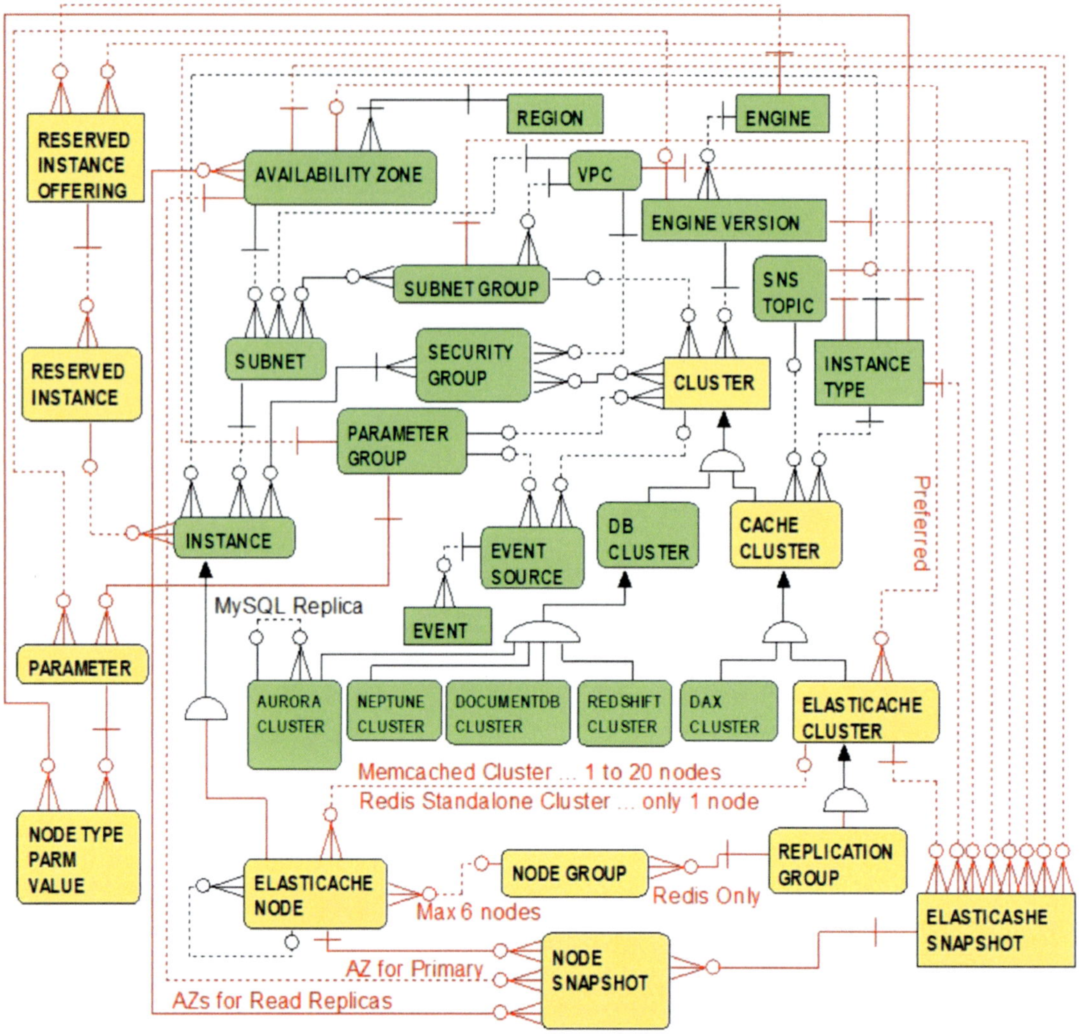

Figure A.9 ElastiCache Basic Concepts

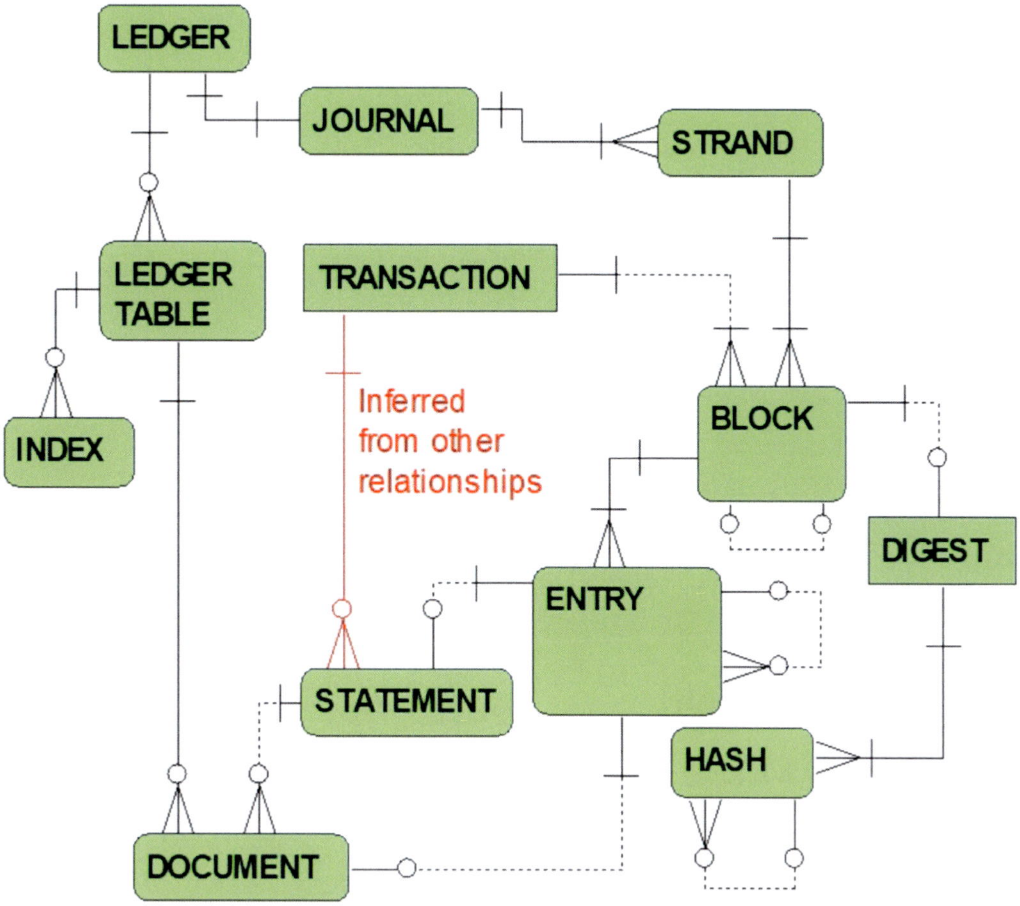

Figure A.10 QLDB Concepts Summary

Chapter References, Notes, Suggested Reading

1. Edgar F. "Ted" Codd - Wikipedia, the free encyclopedia https://en.wikipedia.org/wiki/Edgar_F._Codd

2. Raymond F. Boyce - Wikipedia, the free encyclopedia https://en.wikipedia.org/wiki/Raymond_F._Boyce

3. Donald D. Chamberlin - Wikipedia, the free encyclopedia https://en.wikipedia.org/wiki/Raymond_F._Boyce

4. Amazon database services documents from which the conceptual data model herein is derived can be found at https://docs.aws.amazon.com/ … enabling access to user guides, developer guides, CLI and API references, at the time of this publication under the category "Database."

5. Chapter 1 is adapted from "A Conceptual Data Model for Amazon EC2" … Kindle Edition, by Henry M. Nirsberger, 2017, HMN Consulting, LLC.

6. Recommender Engines - Wikipedia, the free encyclopedia https://en.wikipedia.org/wiki/Recommender_system

7. The NIST Definition of Cloud Computing (SP 800-145) is available at http://csrc.nist.gov/publications/PubsSPs.html#800-145

8. Charles Bachman - Wikipedia, the free encyclopedia https://en.wikipedia.org/wiki/Charles_Bachman

9. *A Short History of the ER Diagram and Information Modeling,* Shannon Kempe, September 25, 2012 http://www.dataversity.net/a-short-history-of-the-er-diagram-and-information-modeling/

10. Simsion, G.C. (2007), *Data Modeling Theory and Practice.* Bradley Beach, NJ: Technics Publications, LLC.

11. *The DAMA Guide to The Data Management Body of Knowledge*, First Edition, Mark Mosely, Editor – Development, Michael Bracket, Editor – Production, Susan Earley, Assistant Editor, Deborah Henderson, Project Sponsor, DAMA International, Technics Publications, LLC, ISBN, print ed. 978-0-9771400-8-4,
    - Page 73 for definition of *data modeling* and *data model.*
    - Page 89 and 90 citing Information Engineering
    - Page 93 for description of *conceptual data model ….*
    - Page 95 for description of *Generalization, Relationships, "a kind of"*
    - Page 98 for description of *Attributes*
    - Page 99 for description of *Keys, Foreign Keys*

12. Simsion, G.C., Witt G.C. (2005). *Data Modeling Essentials.* San Francisco, CA: Morgan Kaufmann Publishers, an imprint of Elsevier.

13. Rumbaugh, J., et.al (1991). *Object-Oriented Modeling and Design*. Englewood Cliffs, NJ: Prentice-Hall, Inc.

14. Martin, J. (1989). *Information Engineering, Book 1: Introduction*. Englewood Cliffs, NJ: Prentice-Hall, Inc.

15. Entity-relationship diagrams herein were produced using IDERA ER/Studio Data Architect, a product of IDERA, Inc. https://www.idera.com

16. Definitions from the AWS Glossary can be found at http://docs.aws.amazon.com/general/latest/gr/glos-chap.html

17. Amazon Simple Storage Service -- Amazon S3 https://aws.amazon.com/s3/

18. Graham, A. (2012). *The Enterprise Data Model: A framework for enterprise data architecture*, Second Edition. Koios Associates Ltd. https://www.amazon.com/Enterprise-Data-Model-enterprise-architecture/dp/0956582915

19. CIDR Notation - Wikipedia, the free encyclopedia for more information on CIDR Blocks, see http://en.wikipedia.org/wiki/CIDR_notation.

20. Amazon Virtual Private Cloud: User Guide (2016) https://aws.amazon.com/documentation/vpc/

21. AWS Key Management Service (KMS) https://aws.amazon.com/kms/

22. Associative entity - Wikipedia, the free encyclopedia https://en.wikipedia.org/wiki/Associative_entity

23. Amazon Simple Notification Service (SNS) https://aws.amazon.com/sns

24. Amazon CloudWatch Logs, https://docs.aws.amazon.com/AmazonCloudWatch/latest/logs/cwl-ug.pdf

25. AWS Identity and Access Management (IAM) https://aws.amazon.com/iam

26. Graph Database - Wikipedia, the free encyclopedia https://en.wikipedia.org/wiki/Graph_database

27. NoSQL - Wikipedia, the free encyclopedia https://en.wikipedia.org/wiki/NoSQL

28. Graph Data Structures - Wikipedia, the free encyclopedia https://en.wikipedia.org/wiki/Graph_(abstract_data_type)

29. GQLs … Gremlin and SPARQL - Wikipedia, the free encyclopedia
https://en.wikipedia.org/wiki/Gremlin_(programming_language)
https://en.wikipedia.org/wiki/SPARQL

30. RDF - Wikipedia, the free encyclopedia
https://en.wikipedia.org/wiki/Resource_Description_Framework

31. Key-Value Pairs (aka Attribute–Value Pairs) - Wikipedia, the free encyclopedia
https://en.wikipedia.org/wiki/Attribute%E2%80%93value_pair

32. Association Class
https://www.ibm.com/support/knowledgecenter/SSCLKU_7.5.5/com.ibm.xtools.modeler.doc/topics/cassnclss.html

33. Labelled Property Graphs (LPGs) - Wikipedia, the free encyclopedia
https://en.wikipedia.org/wiki/Graph_database#Labeled-property_graph

34. Graph Traversal Steps
http://tinkerpop.apache.org/docs/3.3.3/reference/#graph-traversal-steps

35. PRACTICAL GREMLIN: An Apache TinkerPop Tutorial, Kelvin R. Lawrence, version 282-preview, August 28th 2019
https://kelvinlawrence.net/book/Gremlin-Graph-Guide.html

36. Triple Store … http://dbpedia.org/page/Triplestore

37. RDF Reification … https://www.w3.org/TR/rdf-primer/#reification

38. SPARQL commands …. W3C Documentation
https://www.w3.org/TR/rdf-sparql-query/
https://www.w3.org/2008/09/sparql-by-example/Overview.html..

39. RDF4J Console - Wikipedia, the free encyclopedia
https://en.wikipedia.org/wiki/RDF4J

40. JSON-like format … BSON, a binary representation of JSON …
http://bsonspec.org/

41. Agile Software Development - Wikipedia, the free encyclopedia
https://en.wikipedia.org/wiki/Agile_software_development

42. Waterfall vs. Agile methodology …. https://www.seguetech.com/waterfall-vs-agile-methodology/

43. DocumentDB API Vocabulary…
https://docs.mongodb.com/manual/reference/glossary/#term-id

44. MongoDB operations … https://docs.mongodb.com/manual/crud/

45. MongoDB Shell … https://docs.mongodb.com/manual/reference/mongo-shell/

46. SQL Workbench/J … http://www.sql-workbench.eu/

47. Horizontal Partitioning - Wikipedia, the free encyclopedia https://en.wikipedia.org/wiki/Partition_(database)

48. Star Schema - Wikipedia, the free encyclopedia https://en.wikipedia.org/wiki/Star_schema

49. AWS Application Auto Scaling … https://docs.aws.amazon.com/autoscaling/application/userguide/what-is-application-auto-scaling.html

50. Change Data Capture (CDC) - Wikipedia, the free encyclopedia https://en.wikipedia.org/wiki/Change_data_capture

51. Pareto Principle - Wikipedia, the free encyclopedia https://en.wikipedia.org/wiki/Pareto_principle

52. Memcached cache engine - Wikipedia, the free encyclopedia https://en.wikipedia.org/wiki/Memcached

53. Redis cache engine - Wikipedia, the free encyclopedia https://en.wikipedia.org/wiki/Redis

54. Key-Value Store - Wikipedia, the free encyclopedia https://en.m.wikipedia.org/wiki/Key-value_database

55. Lazy Loading caching strategy
    - https://docs.aws.amazon.com/AmazonElastiCache/latest/mem-ug/Strategies.html
    - Wikipedia, the free encyclopedia https://en.wikipedia.org/wiki/Lazy_loading

56. Time to Live - Wikipedia, the free encyclopedia https://en.wikipedia.org/wiki/Time_to_live

57. Write Through caching strategy
    - Wikipedia, the free encyclopedia https://en.wikipedia.org/wiki/Cache_(computing)#Writing_policies
    - WhatIs.com - TechTarget https://whatis.techtarget.com/definition/write-through

58. ElastiCache documentation … https://docs.aws.amazon.com/elasticache/index.html?id=docs_gateway#lang/en_us

59. Data Structures Server … https://redis.io/topics/data-types-intro

60. Examples … Redis data structures … https://www.tutorialspoint.com/redis/index.htm

61. Blob data type - Binary Large Object - Wikipedia, the free encyclopedia https://en.wikipedia.org/wiki/Binary_large_object

62. Binary-Safe data - Wikipedia, the free encyclopedia
https://en.wikipedia.org/wiki/Binary-safe

63. Wikipedia, the free encyclopedia
    - Serialization … https://en.wikipedia.org/wiki/Serialization
    - Base64 … https://en.wikipedia.org/wiki/Base64

64. Partitioning Data Across Redis Nodes … https://redis.io/topics/partitioning

65. *Ledger* - Wikipedia, the free encyclopedia …
https://en.wikipedia.org/wiki/Ledger

66. *Journals* - Wikipedia, the free encyclopedia …
https://en.wikipedia.org/wiki/General_journal

67. System of Record (SOR) applications - Wikipedia, the free encyclopedia …
https://en.wikipedia.org/wiki/System_of_record

68. Separation of Duties (SOD) - Wikipedia, the free encyclopedia …
https://en.wikipedia.org/wiki/Separation_of_duties

69. Amazon Ion … QLDB document format … http://amzn.github.io/ion-docs/docs/spec.html

70. PartiQL … https://partiql.org/

71. *Blockchain* technology - Wikipedia, the free encyclopedia …
https://en.wikipedia.org/wiki/Blockchain

72. SHA-256 … **S**ecure **H**ash **A**lgorithm - Wikipedia, the free encyclopedia …
https://en.wikipedia.org/wiki/SHA-2

73. Cryptographic Hash Function - Wikipedia, the free encyclopedia …
https://en.wikipedia.org/wiki/Cryptographic_hash_function

74. FIPS PUB 180-4, FEDERAL INFORMATION PROCESSING STANDARDS PUBLICATION, Secure Hash Standard (SHS), CATEGORY: COMPUTER SECURITY SUBCATEGORY: CRYPTOGRAPHY .... August 2015, page iv
http://dx.doi.org/10.6028/NIST.FIPS.180-4

75. Merkle Trees - Wikipedia, the free encyclopedia …
https://en.wikipedia.org/wiki/Merkle_tree

76. Merkle Audit Proofs - Wikipedia, the free encyclopedia …
https://www.certificate-transparency.org/log-proofs-work

77. For availability of printed copies of Amazon Database Services summary diagram … email Info@HMNconsulting.com …or visit www.HMNconsulting.com.

Made in United States
Orlando, FL
04 November 2026

90200295R00117